Extension **Mathematics**

For Higher Achieving Students

Lower Secondary

Student Edition

First published in Australia in 2009
Phoenix Education Pty Ltd
PO Box 3141
Putney NSW 2112
Australia

Phone	02 9809 3579
Fax	02 9808 1430
Email	service@phoenixeduc.com
Web	www.phoenixeduc.com

Extension Mathematics for Higher Achieving Students
Junior Secondary, Student Edtion

ISBN 978 1 921085 80 2

Printed in Australia by Five Senses Education

Contents

Chapter 8
Statistics

Chapter 9
Probability

Chapter 10
Problem Solving

Introduction

Thinking and working mathematically

In a rapidly changing world, rather than acquiring facts and knowledge it is the ability to think effectively and solve problems in a variety of situations that will serve future generations well. Students need to be encouraged to think creatively, to be flexible in the use of new ideas and approaches, to be fluent in the creation of an array of ideas, to elaborate on similar ideas and to be original in thinking of new approaches or new ways of seeing problems.

Aim of resource

The broad aim of this resource is to provide students, teachers and parents with a range of problems and tasks centred largely on the topic and concept areas studied in the upper middle curriculum. The problems and tasks also have points of departure to allow challenge, application and extension of ideas.

The solving of these problems requires each student to seek information outside the instruction delivered in the classroom framework, to modify their thinking and be prepared to use and think of new approaches in unfamiliar situations. Where appropriate, skills and formulae of mathematics that will be covered in later years is included, which will broaden the student's mathematical toolbox.

Rather than teach students how to think strategically, the problems contained in this book encourage the application of thinking skills as being a natural part of the process. Students working in groups can share thinking and ideas, from which each individual can learn.

From a teaching point-of-view, it is crucial that students be required to document what they have learnt and what thinking was employed to arrive at solutions to the tasks. Students will be encouraged to value the solution pathway as much, if not more than that of the answer.

Principles affirmed by a range of national bodies

The following are quotes from Australian and International organisations that outline guiding principles for education from a mathematical perspective. This resource provides educational institutions with material that can be used to affirm international expectations of mathematics education.

The AAMT (Australian Association of Mathematics Teachers) standards, 2005 indicate that excellent teachers:

- arouse curiosity, challenge student's thinking and engage them actively in learning
- negotiate mathematical meaning, model mathematical thinking and reasoning
- promote, expect and support creative thinking, mathematical risk-taking in finding and explaining solutions.

The AAMTC "Mathematical knowledge and understanding for effective participation in Australian Society, 1996".
The development of problem solving strategies requires that students have many opportunities to solve a wide variety of problems. Such activities provide opportunities for students to engage in individual and cooperative work and to extend their mathematical thinking through problem posing, reflection and persistence with difficult tasks.
Students need experiences in applying mathematics to practical problems, in acquiring a variety of standard mathematical techniques and choosing appropriately between them, in using non-routine applications, and in using mathematical modelling techniques to solve problems.
(www.aamt.edu.au)

The National Council of Teachers of Mathematics (NCTM) – USA, proposed, "Problem solving should be the central focus of mathematics education. Mathematics should be seen as being a body of knowledge and a way of thinking that is useful in approaching problems encountered in everyday situations. To promote higher-order thinking, more problems should be confronted for which there are alternative solution strategies and solutions to generate and debate". (www.standards.nctm.org).

Fostering and encouraging mathematical thinking for high-achieving students

Mathematics is a language: a structure based on logical thought with number being the central component. Finding patterns and generalising numerical solutions to problems beyond a limited number of observations leads to abstraction with the use of algebraic notation. Mathematical models such as graphs, the unit circle for trigonometry, Venn diagrams for probability, box-and-whisker plots for statistics provide a visual profile that can be read and interpreted. Calculators and technology play an increasingly important role in the elimination of the drudgery of numerical calculation as well as providing a range of visual representations. Proof and justification of valid solution steps is crucial in charting a route from question to answer. Geometric facts are always physically true, and to be used they need to be known such as when working with complex patterns of polygons to find unknown angles. The history of mathematics allows us to put our current knowledge into a framework of discovery such as the transition from Roman numerals to Arabic numerals to the establishment of imaginary numbers, which do not physically exist. The recognition of axioms – mathematical truths or laws – that underpin all mathematical argument often take students full circle, where all solutions are questionable.

The high-achieving mathematical student

The mathematically talented student can range in working method from instinctive, intuitive thinker through to the slow, methodical worker. Either way it is the sophistication and maturity of thought, the perseverance to achieve clarity, the breakthrough to find initial solutions and, finally, the will to go on and open up the task to lead to further investigation, asking "what would happen if ..." or " is there a better solution or set of solutions", that is to be encouraged. It is not the answer that is the important outcome for many of the tasks in this book but rather the thinking pathway that can open doors to further intellectual exploration.

Students who are talented in mathematics when presented with a task or problem are able to:

- feel confident and positive in the challenge of approaching a new problem
- understand quickly the nature and context of the question
- assemble a collection of concepts or approaches that could be helpful
- make connections between similar situations that have previously been solved successfully
- think their way around the problem and look at it from different viewpoints
- choose from a number of models the one or ones that might be useful
- play with the problem in their mind and make rough sketches before completing the finished product.

Some students think quicker than they can record their working. They must be encouraged to record their work in jottings as they speed through the process but they must provide a detailed account of the full solution. I also encourage students to list what they have learnt at the end of each problem – skills, approach, methodology and enjoyment.

Different learning and thinking styles

Students have different ways of perceiving the world and apply different thinking approaches when presented with unfamiliar problems.

Visual thinkers process problems in a physical sense developing pictures or representations that help them develop a three-dimensional appreciation of the problem. For these students, geometry is trivially simple – the answer stares them in the face with an obviousness that makes explanation pointless. Students in this category must be challenged to explain and quantify what they see rather than stating the answer in a self-evident fashion.

Algebraic thinkers leap into abstract thought and perceive things in terms of variables and constants, and then endeavour to find the relationships between them. They work with values and dimensions and understand the restrictions or constraints that could or do apply to a situation.

Using the resource

Dynamics of student groups

When schools decide to use resources such as this one and accommodate high achieving students in their mathematics program, parents should be informed. It is a bonus that the school makes available to their child. The easiest and most direct way is to require parents to sign each piece of work when it is completed. In this way, students have to explain what it is that their parents are signing. Articles in the school newsletter profiling the program along with the names of students who have achieved notable achievement are positive ways of keeping the school community informed.

Using the resource in the classroom

Use individual questions in class when students finish their work as the sheets have been grouped into topic areas.

Students who have demonstrated mastery of the concepts used in a topic can work through the extension sheets instead of working through sets of boring exercises.

Alternatively, give a sheet of questions to the class at the end of a topic to extend the group. Students can work individually or in groups according to their preferred method of working. Weaker students can be given the set of hints, whereas others can work on unaided. Marks, points or edible rewards can be awarded to motivate interest.

Using the resource in maths extension/enrichment groups

Groups of like-minded students can meet at lunchtime or before school. These groups provide students with the opportunity to mix with others of similar ability from other classes and so broaden their friendship groups. Students can engage freely in healthy, challenging debate. Tokens of recognition such as badges and certificates would be beneficial to encourage participation. Students could also receive suitable recognition in their school report.

Using the resource in a homework program

High achieving students could be given selected questions from a sheet rather than the standard homework sheet. The timeframe for the solution of the problems can be negotiated.

Hints section in the teacher edition

Students sometimes prefer to work individually or in groups to pool ideas when a challenging problem is presented but it can be unfair to simply give out problems and expect students to connect with them without some initial guidance. A hints section is provided, question-by-question for each sheet except for the challenging section, which is non-routine, so that when it is needed students can be given some insight as to the structure and context for each task. Often students are blocked from answering seemingly difficult questions by not recognising or not knowing basic definitions. The hints section provides an insight into the structure and meaning of each question ranging from what to look for to pointing out some key initial steps.

Often tasks are constructed according to a set routine method of solution; however, it is a noteworthy event when students can find different solution pathways to arrive at the same or multiple valid answers. These alternative workings should be collected and discussed with the group. I have provided solutions for all tasks as a guide for parents and teachers who need to find a quick profile of each problem. These are contained in the *Teacher Edition*

Chapter 1

Number

Areas of Interest

Cross-number Puzzle (1) to (4)
Arithmetic Challenge
Fractions
Directed Number (1) & (2)
Ratios (1) & (2)
Percentages (1) & (2)
Consumer Arithmetic (1) & (2)
Prime Numbers
Factors
Multiples and Factors
Numbers and Algebra
Numbers and Indices (1) & (2)
Numbers and Square Roots
Number Patterns (1) & (2)

Cross-number Puzzle (1)

Use the clues to complete the cross-number puzzle.

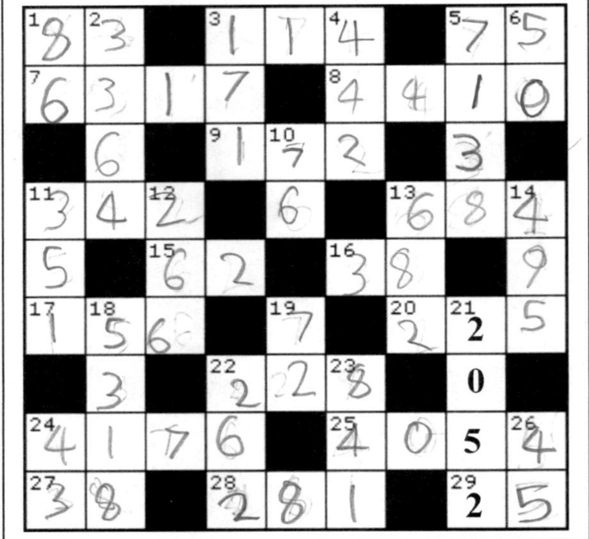

Across

3. *27 across* times three
5. *6 down* plus twenty-five
7. *5 down* minus 821
8. *2 down* plus 1046
9. *24 down* times four
11. *21 down* divided by six
13. *21 down* divided by three
15. *13 down* divided by eleven
16. *12 down* divided by seven
17. *9 across* minus sixteen
20. A square number
22. *20 across* plus three
24. *8 across* minus 234
25. *24 across* minus 122
27. *22 across* divided by six
28. *22 down* plus nineteen
29. *26 down* minus twenty; also a square number

Down

2. *23 down* times four
3. *21 down* divided by twelve
4. *28 across* plus 161
5. *1 across* times *1 down*
6. *29 across* times two
10. *13 across* divided by nine
11. *3 down* plus 180
12. Seven times *27 across*
13. *23 down* minus 159
14. *26 down* times eleven
18. *8 across* plus 908
19. *10 down* minus four
21. *22 across* times nine
22. *22 across* plus thirty-four
23. A square number

Cross-number Puzzle (2)

Use the clues to complete the cross-number puzzle.

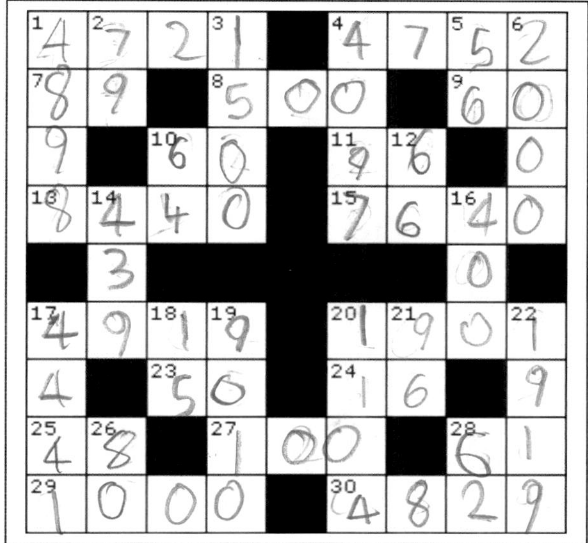

Across

1. *1 down* minus 177
4. *1 across* plus thirty-one
7. *9 across* plus twenty-nine
8. Five times *27 across*
9. Four times *18 down*
10. *11 across* minus thirty-six
11. *5 down* plus forty
13. *17 across* plus 3521
15. *13 across* minus 800
17. *1 down* plus twenty-one
20. *6 down* minus ninety-nine
23. *16 down* divided by eight
24. A square number
25. *24 across* times three
27. A square number
28. 18 *down* plus 46
29. Two times *8 across*
30. *17 across* minus ninety

Down

1. *2 down* times *28 down*
2. Two less than a square number
3. Three times *8 across*
4. *1 across* minus 624
5. *2 down* minus twenty-three
6. Two times *29 across*
10. Four times *24 across*
12. *25 across* plus eighteen
14. *30 across* divided by eleven
16. *6 down* divided by five
17. *1 across* minus 280
18. *23 across* minus thirty-five
19. *13 across* plus 570
20. *29 across* plus 104
21. Six times *24 across*
22. *20 across* plus eighteen
26. *10 down* plus sixteen
28. *28 across* plus one

Hint: Start with 24 *across* and 21 *down* to start the puzzle.

Cross-number Puzzle (3)

1. Find the values of a and b to complete the cross-number puzzle.

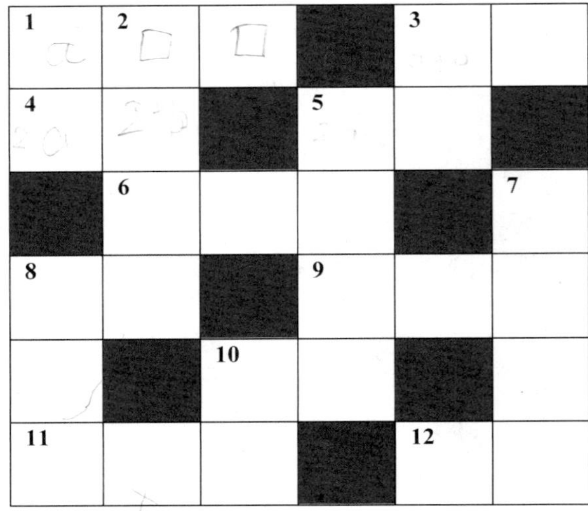

Across
1. a^2
3. $a+b$
4. $2b$
5. $2a$
6. $45a-3$
8. b
9. $a(a+b)$
10. $\left(\frac{1}{2}a\right)^2+45$
11. b^2
12. $b+2a$

Down
1. a
2. b^3
3. $2(b-2)$
5. a^3
7. b^3-3a^3
8. ab
10. $5b-a$

2. Use the clues to complete this challenging cross-number puzzle.

1	2	3		4	
5					
			6		7
		8			
9				10	
		11			

Across
1. $9\times(2\ down)$
4. A square number
8. The same as $(10\ down)+(10\ across)$
and $9\times(6\ down)$
9. Divisible by 11

Down
1. $78\times(3\ down)$
2. $9\times(3\ down)$
3. Both digits are the same
4. One more than $(2\ down)$
7. $9\times\ (6\ across)$
8. $90+\ (11\ across)$
and all the digits are the same
9. $3\times(3\ down)$

Hint 1: Start with ***2 down*** and ***3 down*** where nine times a two-digit number equals a two-digit number – don't forget that ***3 down*** is special.
Hint 2: Look at ***8 across*** and ***6 down*** – both are two-digit numbers that end with the same digit and ***8 across*** = 9 × ***6 down.***

Cross-number Puzzle (4)

1. The remaining numbers in the two puzzles are square numbers. Find the numbers and write the clues.

(a)

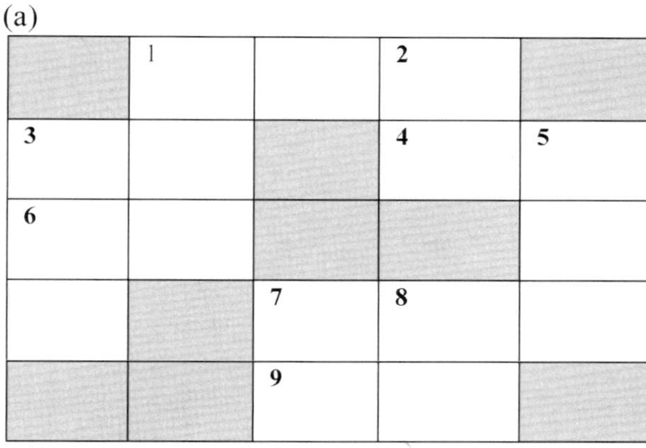

Across
1
3
4
6
7
9

Down
1
2
3 *a cube number*
5
7
8

(b)

	1		2	
3			4	5
6				
		7	8	
	9			

Across
1
3
4
6
7
9

Down
1
2
3 *a cube number*
5
7
8

2. The numbers in the puzzle are all squares. Find the numbers and write the clues to show all possible solutions.

	1	2	3	
4				5
6	1		7	
8		9		1
	10			

Across
1
4
6
7
8
9
10

Down
1
2
3
4
5
7
9

Arithmetic Challenge

1. Each pronumeral stands for a digit in a number. Find the value for the pronumerals.

(a)
$$\begin{array}{r} 2\,4\,a\,9\,b \\ c\,d\,7\,2\,6\;+ \\ \hline 4\,1\,6\,e\,1 \\ \hline \end{array}$$

(b)
$$\begin{array}{r} 2\,a\,6\,7 \\ 5\,b\,3 \\ c\,4\,6\,d\;+ \\ \hline 5\,7\,1\,4 \\ \hline \end{array}$$

(c)
$$\begin{array}{r} d\,2\,b\,1\,e \\ 5\,c\,5\,d\,3\;- \\ \hline 3\,1\,4\,9\,6 \\ \hline \end{array}$$

(d)
$$\begin{array}{r} 3\,a \\ \times 4\,2 \\ \hline 7\,6 \\ b\,c\,d\,0\;+ \\ \hline 1\,e\,f\,6 \\ \hline \end{array}$$

(e)
$$\begin{array}{r} 2\,a\,b \\ \times\,6\,7 \\ \hline 2\,0\,8\,6 \\ 1\,c\,d\,e\,0\;+ \\ \hline 1\,f\,g\,h\,6 \\ \hline \end{array}$$

(f)
$$\begin{array}{r} a\,b\,7\,6\,4 \\ 7\,\overline{)\,8\,2\,c\,4\,d} \end{array}$$

(g)
$$\begin{array}{r} 2\,6\,0\,0\,3 \\ a\,\overline{)\,7\,8\,b\,c\,d} \end{array}$$

2. Each pronumeral stands for a digit in a number. Find the value of the pronumerals.

(a)
$$\begin{array}{r} a\,a\,b\,a \\ b\,8\,2\;+ \\ \hline a\,9\,a\,b \\ \hline \end{array}$$

(b)
$$\begin{array}{r} a\,b\,a \\ a\,a \\ b\,9\,b\;+ \\ \hline 1\,a\,6\,2 \\ \hline \end{array}$$

(c)
$$\begin{array}{r} a\,b\,a \\ b\,b\,a\,a \\ c\,c\,a\;+ \\ \hline 5\,7\,6\,c \\ \hline \end{array}$$

3. Each pronumeral stands for a digit in a number. Find the values of the pronumerals – there is more than one answer.

(a)
$$\begin{array}{r} 9\,a\,2\,b\,3 \\ c\,6\,8\,d\;+ \\ \hline e\,2\,f\,5\,2 \\ \hline \end{array}$$

(b)
$$\begin{array}{r} 2\,a \\ b\,c\,8 \\ 8\,0\,3\,d\;+ \\ \hline 9\,e\,0\,1 \\ \hline \end{array}$$

(c)
$$\begin{array}{r} a\,b\,a \\ \times\,b\,b \\ \hline a\,b\,d \\ a\,b\,a\,0\;+ \\ \hline a\,c\,c\,a \\ \hline \end{array}$$

(d)
$$\begin{array}{r} a\,a\,a \\ \times\,b\,b \\ \hline b\,b\,b \\ b\,b\,b\,0\;+ \\ \hline b\,c\,c\,b \\ \hline \end{array}$$

(e)
$$\begin{array}{r} a\,0\,b \\ 5\,c\,3\;- \\ \hline 2\,8\,d \\ \hline \end{array}$$

(f)
$$\begin{array}{r} 3\,6\,a\,b \\ c\,6\,5\,4\;- \\ \hline d\,e\,f\,8 \\ \hline \end{array}$$

Fractions

1. (a) Calculate the following.

(i) $\left(1-\frac{1}{2}\right)\left(1-\frac{1}{3}\right)$

(ii) $\left(1-\frac{1}{2}\right)\left(1-\frac{1}{3}\right)\left(1-\frac{1}{4}\right)$

(iii) $\left(1-\frac{1}{2}\right)\left(1-\frac{1}{3}\right)\left(1-\frac{1}{4}\right)\left(1-\frac{1}{5}\right)$

(iv) If n is a positive integer express $\left(1-\frac{1}{2}\right)\left(1-\frac{1}{3}\right)\left(1-\frac{1}{4}\right)\left(1-\frac{1}{5}\right)....\left(1-\frac{1}{n}\right)$ in terms of n.

(b) Calculate the following.

(i) $\left(1+\frac{1}{2}\right)\left(1+\frac{1}{3}\right)$

(ii) $\left(1+\frac{1}{2}\right)\left(1+\frac{1}{3}\right)\left(1+\frac{1}{4}\right)$

(iii) $\left(1+\frac{1}{2}\right)\left(1+\frac{1}{3}\right)\left(1+\frac{1}{4}\right)\left(1+\frac{1}{5}\right)$

(iv) If n is a positive integer express $\left(1+\frac{1}{2}\right)\left(1+\frac{1}{3}\right)\left(1+\frac{1}{4}\right)\left(1+\frac{1}{5}\right)....\left(1+\frac{1}{n}\right)$ in terms of n.

(c) Calculate the following.

(i) $\left(1-\frac{1}{2}\right)\left(1+\frac{1}{3}\right)$

(ii) $\left(1-\frac{1}{2}\right)\left(1+\frac{1}{3}\right)\left(1-\frac{1}{4}\right)$

(iii) $\left(1-\frac{1}{2}\right)\left(1+\frac{1}{3}\right)\left(1-\frac{1}{4}\right)\left(1+\frac{1}{5}\right)$

(iv) $\left(1-\frac{1}{2}\right)\left(1+\frac{1}{3}\right)\left(1-\frac{1}{4}\right)\left(1+\frac{1}{5}\right)\left(1-\frac{1}{6}\right)$

Fractions page 2

1. (c) (v) For the expression $\left(1-\frac{1}{2}\right)\left(1+\frac{1}{3}\right)\left(1-\frac{1}{4}\right)\left(1+\frac{1}{5}\right).....\left(1+\frac{1}{n}\right)$, show that there is an even number of terms and hence express it in terms of n.

(vi) Find the value of $\left(1-\frac{1}{2}\right)\left(1+\frac{1}{3}\right)\left(1-\frac{1}{4}\right)\left(1+\frac{1}{5}\right).....\left(1-\frac{1}{n}\right)$ and define the values for n.

2. Evaluate the following.

(a) $2+\cfrac{1}{2+\cfrac{1}{2+\frac{1}{2}}}$

(b) $2-\cfrac{1}{2-\cfrac{1}{2-\frac{1}{2}}}$

(c) $2+\cfrac{1}{2+\cfrac{1}{2+\cfrac{1}{2+\frac{1}{2}}}}$

(d) $2-\cfrac{1}{2-\cfrac{1}{2-\cfrac{1}{2-\frac{1}{2}}}}$

Directed Number (1)

1. Four coloured discs have letters and numbers printed on them so that $a = 1,\ b = -2,\ c = 7,\ d = -11$.

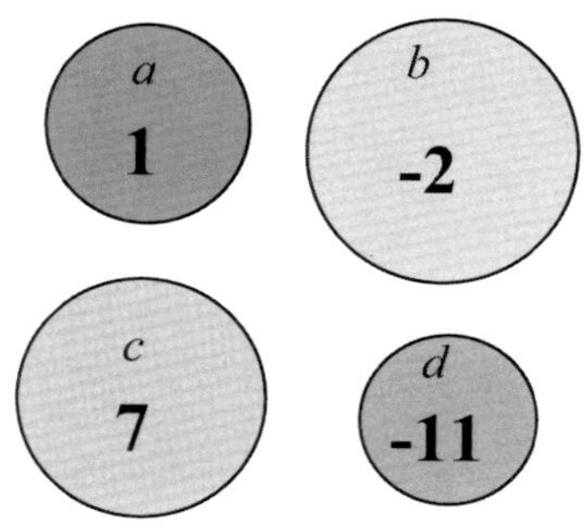

The discs can be combined to make different totals. A total of zero can be made by using three a's, one b, three c's and two d's. This can be calculated by:

$3a + b + 3c + 2d$

$= 3 \times 1 + 1 \times -2 + 3 \times 7 + 2 \times -11$

$= 3 - 2 + 21 - 22 = 0$

Use discs of each type, but no more than three discs of any type, to make totals of $\{3, 2, 1, 0, ..., -4, -5, -6\}$; find another combination to make zero!

a+a+a=3
a+a=2
a=1 = b+a+a+a=1   b+a+a=0   b+b=-4

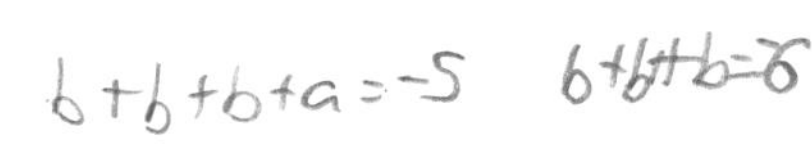

2. Four coloured discs have letters and numbers printed on them so that $a = 2,\ b = -3,\ c = 6,\ d = -1$. The product of a and b added to the product of c and d can be found:

$ab + cd$

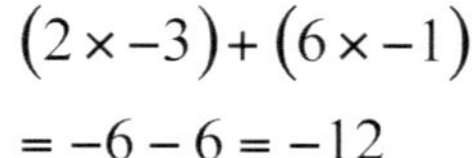

$(2 \times -3) + (6 \times -1)$

$= -6 - 6 = -12$

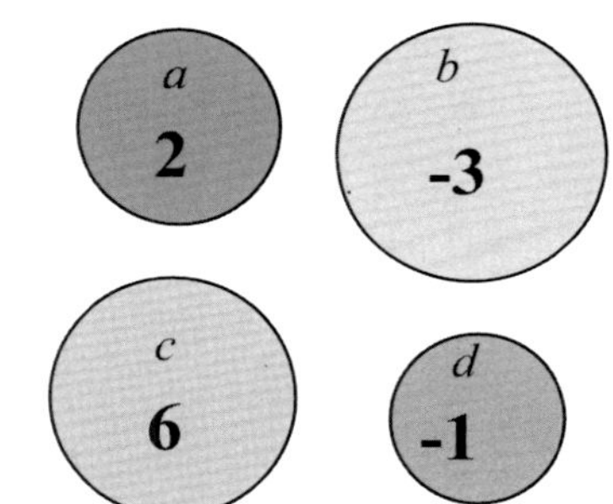

(a) Find the sum of two products of the numbers on the discs to make the totals of:

(i) 15
a+a+c+c+d=15
2+2+6+6-1=15

(ii) −20

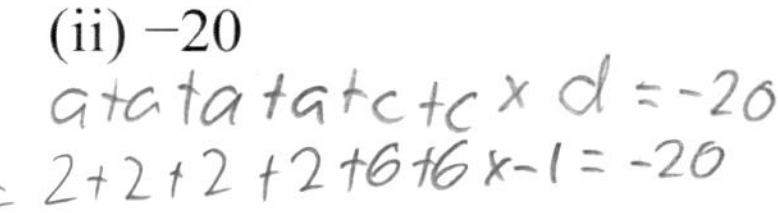

(b) Find the sum of the product of the numbers on two discs with another disc to make:

(i) −7 b+b+d=-7
-3+-3+-1=-7

(ii) 4 a+a=4
2+2=4

(iii) 5 c-d=5
6-1=5

(iv) −4 b+d=-4
-3+-1=-4

(v) 9 a+a+c-d=9
2+2+6-1=9

(vi) −9 b×b=-9
-3×3=-9

(c) Find all the possible results when the product of the numbers on three discs is added to the number on the remaining disc.

Directed Number (1) page 2

3. Coloured tiles with consecutive negative numbers −6, −7, −8 and −9 are placed on a board with + or – operation discs between them.

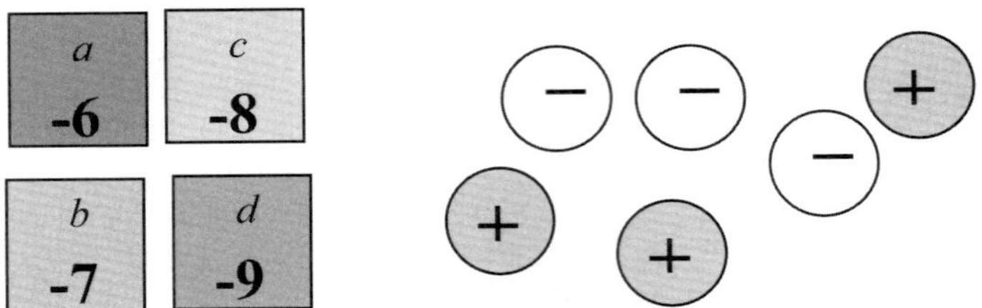

One way that this can happen is:

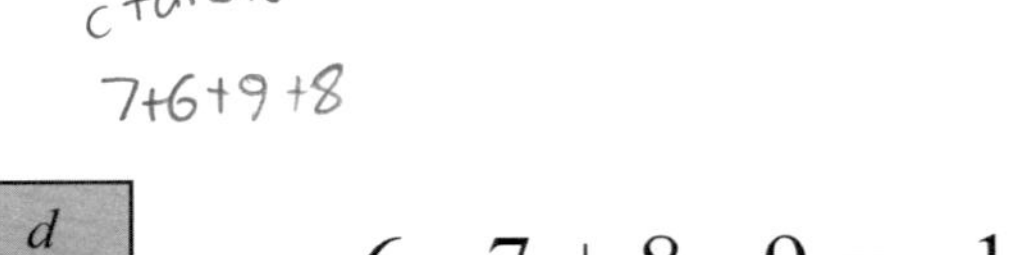

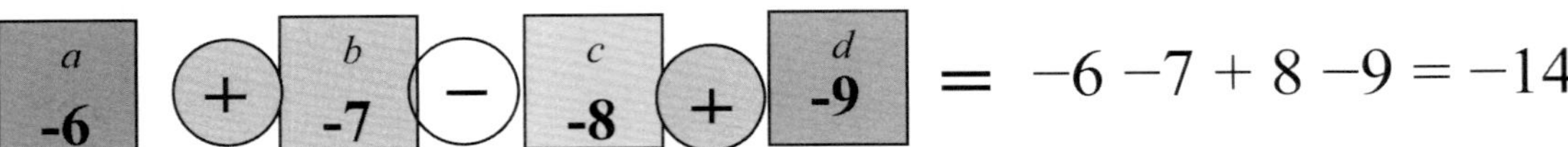

List all the possible answers when the operation discs are placed in front of and between the number tiles.

Directed Number (2)

1. The numbers on the faces of a die are $\{-1, -2, -3, -4, -5, -6\}$.

   (a) Place the numbers on the die so that the opposite faces add to the same sum.

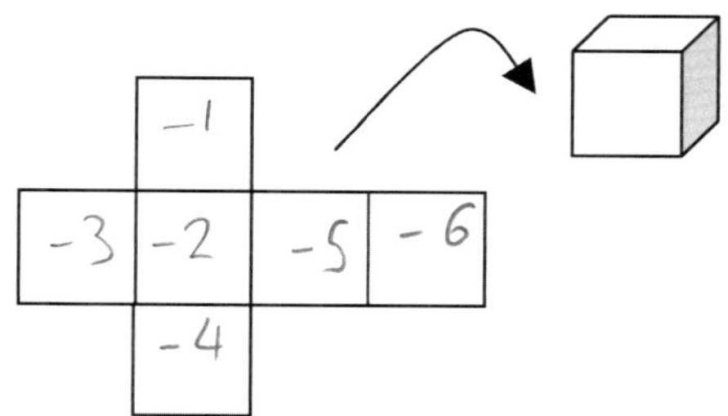

   (b) If the dice is rolled twice, find the probability that the numbers:

   (i) add to an even number

   $\frac{1}{2}$

   (ii) subtract to a positive odd number $\frac{3}{6}$

   (iii) subtract to a negative even number

   $\frac{1}{3}$

   (iii) multiply to make an even positive number.

   $\frac{6}{9}$

   3729634

2. (a) Find all possible sums of two integers with a product of:

   (i) −6

   -3 -2

   -1 -6

   (ii) 12

   6, 2, 1, 12

   , 3, 4

   (b) Find the products, using three different integers for the following, and find the sum of the integers for each set.

   (i) 24

   6, 4, 12, 9, 2, 3, 4

   (ii) −36

   -2, -18, -12, -4, -3, -19, - 20

Directed Number (2) page 2

3. Consecutive numbers are separated by one unit such as 2, 3, 4, ….
   (a) Use the four consecutive numbers 2, 3, 4 and 5 to calculate the following.
   (i) $2 + 3 + 4 + 5 =$ ___ (ii) $2 - 3 + 4 + 5 =$ ___

   (b) Place + or – signs in the spaces shown for the remaining 14 combinations and calculate the answers.

___ 2 ___ 3 ___ 4 ___ 5 =	___ 2 ___ 3 ___ 4 ___ 5 =	___ 2 ___ 3 ___ 4 ___ 5 =
___ 2 ___ 3 ___ 4 ___ 5 =	___ 2 ___ 3 ___ 4 ___ 5 =	___ 2 ___ 3 ___ 4 ___ 5 =
___ 2 ___ 3 ___ 4 ___ 5 =	___ 2 ___ 3 ___ 4 ___ 5 =	___ 2 ___ 3 ___ 4 ___ 5 =
___ 2 ___ 3 ___ 4 ___ 5 =	___ 2 ___ 3 ___ 4 ___ 5 =	___ 2 ___ 3 ___ 4 ___ 5 =
___ 2 ___ 3 ___ 4 ___ 5 =		

   (c) Repeat the process for the general set of 4 consecutive numbers $n,\ (n+1),\ (n+2),\ (n+3)$ and make some conclusions comparing the two sets of numbers.

4. (a) Use the consecutive negative numbers −2, −3, −4 and −5 to calculate the following.
   (i) $-2 + -3 + -4 + -5 =$ ___ (ii) $-2 \ - -3 + -4 + -5 =$ ___

   (b) Place + or – signs between the numbers for the remaining 6 combinations and calculate the answers.

−2 ___ −3 ___ −4 ___ −5 =	−2 ___ −3 ___ −4 ___ −5 =	−2 ___ −3 ___ −4 ___ −5 =
−2 ___ −3 ___ −4 ___ −5 =	−2 ___ −3 ___ −4 ___ −5 =	−2 ___ −3 ___ −4 ___ −5 =

   (c) Repeat the process for the general set of 4 consecutive numbers $-n,\ -(n+1),\ -(n+2),\ -(n+3)$ and make some conclusions comparing the two sets of numbers.

Ratios (1)

1. Find the ratio of $y{:}z$ if:

   (a) $2x = 3y - 8$ and $x + 4 = 10z$ 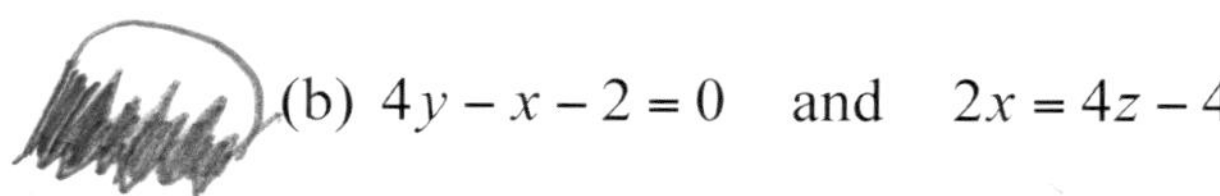 (b) $4y - x - 2 = 0$ and $2x = 4z - 4$

2. The dimensions of a rectangular box are in the ratio of 2:3:5 and its volume is 147 390cm^3.

   (a) Find the dimensions of the box.

   (b) Find the ratio of the surface area of the three different sized faces.

3. Two 240mL bottles contain water and orange concentrate in the ratio 4:1 and 2:1 respectively.

   (a) Find the amount of water and orange concentrate in each bottle.

   (b) (i) If the contents of the two are mixed in a large container, find the new ratio of water to orange concentrate.

   (ii) Some liquid is removed from the container and replaced with the same amount of water. Find the volume removed and then replaced so that the ratio of water to concentrate is 3:1.

Ratios (1) page 2

4. The ratio of girls to boys in a tennis club is 2:3. When ten more girls and five more boys joined the club, the ratio is now 3:4.

(a) How many members were in the club originally?

(b)

(b) How many members are in the club now?

5. The shape is made up of semicircles with centre •. The area of the shapes marked *C* are the same as each other and the shapes marked *B* have area as each other.

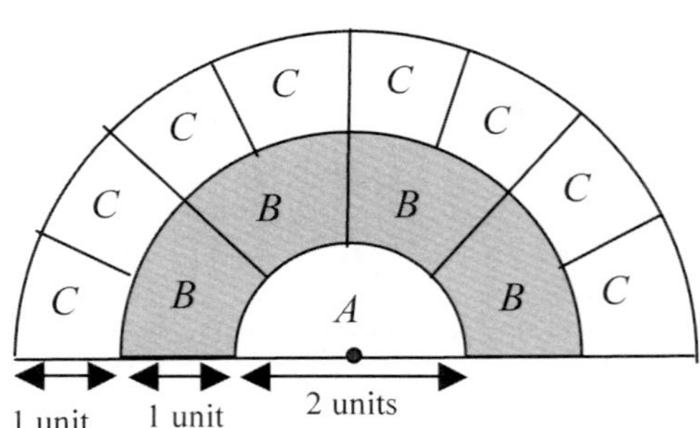

(a) Find the ratio of the area of the shapes $A:B:C$.

(b) Find the ratio of the perimeter of the shapes $A:B:C$.

Ratios (2)

1. A shaded square is placed inside a larger square as shown.
   (a) Find the ratio of the area of the shaded square to the area of the larger square.

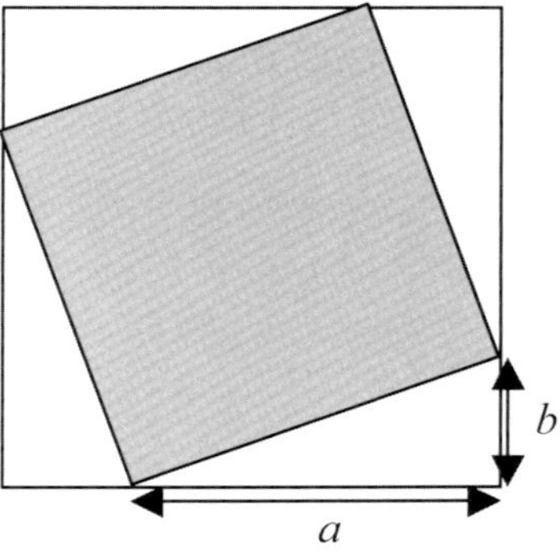

   (b) Find an expression for the edge length of the shaded square in terms of a and b and hence, find the ratio of the perimeter of the shaded square to the perimeter of the larger square.

   (c) Find the relationship between a and b for the ratio of the area of the shaded square: the area of the larger square = 1:2.

2. Two numbers are in the ratio 5:2. If the difference between them is 57, find:
   (a) their sum
   (b) their product.

3. Two cylinders of concrete have their base diameters in the ratio 2:3 and their heights in the ratio 2:1, respectively.
   Find the ratio of their:
   (a) volumes
   (b) surface areas.

Ratios (2) page 2

4. The whole numbers a, b and c are in the ratios $a:b = 2:3$ and $a:(b+c) = 3:5$.

(a) Find the ratios:

(i) $a:c$

(ii) $b:c$

(b) Find the first three positive whole number sets of values of a, b and c.

5. Red and blue marbles are placed in a bag in the ratio $a:a+2$. Find the number of marbles that need to be added to the bag so that the ratio of red to blue marbles becomes:

(a) 1:1

(b) $a:a+4$

(c) 1:2

(d) $a+2:a$

Percentages (1)

1. A basketball team won 28 out of 42 games played and has 18 still to play. How many of the remaining games must it win in order that its percentage of games won for the entire season will be 75%? 45 games must be won.

2. (a) 65% of a number is 15, find the number.

   (b) 30% of a number is one-thirtieth of x, find the initial expression as a fraction of x.

   $\frac{1}{9}$ $\frac{x}{9}$

3. All the questions set in an examination are of equal value. A student answers 20 out of the first 32 questions correctly, but gets three-quarters of the remaining questions wrong. If this student's score is 49%, how many questions were set in the examination? 63

4. William and Lucinda are given the same number of badges to sell at the local fete. William sells 90% of his badges, while Lucinda sells $\frac{3}{4}$ of her badges. If between them they raise \$495, how much more money has William raised compared to Lucinda? \$193 more dollars

Percentages (1) page 2

5. Orange juice concentrate, which is 25% water, is imported.
   (a) How many litres of water must be added to a 100-litre container of the orange juice concentrate to produce a mixture that is $\frac{1}{3}$ water?

33.3̇ %

   (b) A twelve-litre container of the orange juice concentrate has 4 litres removed. The container is filled up with water, mixed thoroughly and a further 4 litres of this mixture is then removed. If the container is filled up with water again, find what percentage of the final mixture is water. 44.

6. A man buys a house for \$650 000. He borrows \$350 000 from the bank, paying 11% interest per annum on an investment loan. He leases the house to a tenant who pays rent of \$680 per week less 8% management fee charged by a real estate agent. If the owner pays \$4 500 on rates and taxes and \$4 000 on repairs, find:
   (a) his total expenses 918 500

   (b) his total return from the investment

1100

   (c) the percentage return from his investment.

1%

   (d) Comment on the strategy of this investment versus investing the capital in a bank investment loan of 8%.

???

Percentages (2)

1. (a) Find the percentage change to a number if it is increased then decreased by:
   (i) 10% (ii) 20% (iii) 40%

   (b) A number is increased then decreased by a certain percentage. Find the percentage applied if the number is decreased by:
   (i) 9% (ii) 25% (iii) 36%

   (c) Find the relationship between the overall percentage change and the percentage increase then decrease applied to the number.

2. Find the percentage change to the area of a square after its sides are:
   (a) increased by 10% (b) decreased by 10%

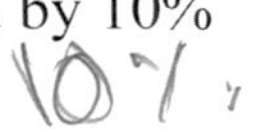

   (c) doubled

   (d) halved.

3. (a) Find the percentage change to the area of a rectangle after:
   (i) its length is doubled and its width halved

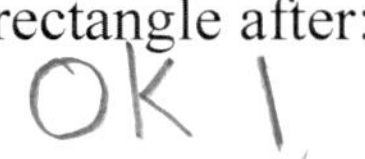

   (ii) its length is increased by 20% and its width is decreased by 10%.

   (b) A rectangle has its length increased by 25%. Find the percentage change to the width so that the area of the rectangle remains unchanged.

Percentages (2) page 2

4. A square section was cut from the face of a cube where the lengths marked with dashes are the same length.

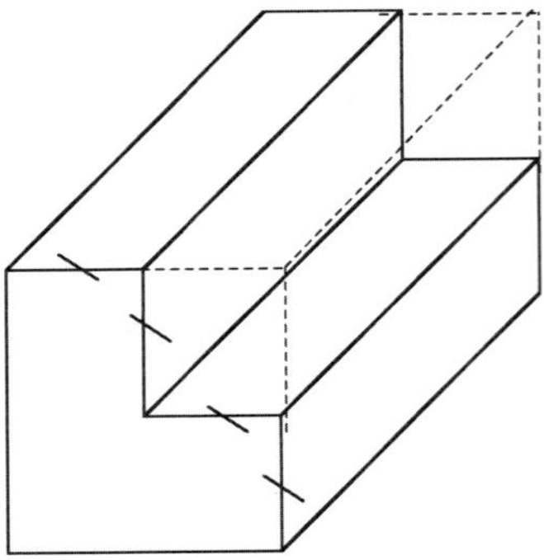

When the square section is removed, find the percentage change to the:
(a) length of the edges 25%

(b) surface area 25%

(c) volume. 25%

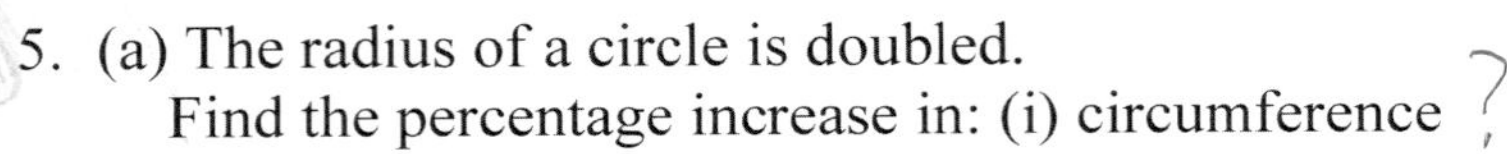

5. (a) The radius of a circle is doubled.
Find the percentage increase in: (i) circumference ? (ii) area. ?

(b) The radius, r, of a sphere is doubled.
(i) Find the percentage increase to the surface area if the surface area of a sphere is $SA_{sphere} = 4\pi r^2$. ?

(ii) Find the percentage increase in volume given that the volume of a sphere is $V_{sphere} = \frac{4\pi r^3}{3}$. ?

Consumer Arithmetic (1)

The income tax scale for Australian residents is shown below.

Tax rates 2007– 08

Taxable income	*Tax on this income*
\$1 – \$6 000	Nil
\$6 001 – \$30 000	15c for each \$1 over \$6 000
\$30 001 – \$75 000	\$3 600 plus 30c for each \$1 over \$30 000
\$75 001 – \$150 000	\$17 100 plus 40c for each \$1 over \$75 000
\$150 001 and over	\$47 100 plus 45c for each \$1 over \$150 000

The Medicare levy is payable at the rate of 1.5% of people's taxable income unless they are exempt for reasons such as being entitled to full free medical treatment for all conditions under defence force arrangements or Veterans' Affairs Repatriation Health Card (Gold Card) or repatriation arrangements.

The Medicare levy surcharge is payable by individuals who do not have private health insurance at the rate of 1% of their taxable income and who are above the threshold: single people without dependants who earn more than \$50 000, or families with combined income according to the table:

Number of dependent children	**Surcharge income threshold**
0–1	\$100 000
2	\$101 500
3	\$103 000
4	\$104 500
More than 4 dependent children	\$104 500 plus \$1 500 for each additional child

Consumer Arithmetic (1) page 2

1. Use the information above to calculate the tax payable by the following people.
   (a) Sam is single and does not have private health insurance. His taxable income is \$52 543 for the year.

   (b) Huy works at the rate of \$21.50 per hour for a fifty-hour week and his Amy wife earns \$58 600 per year. They have one child but do not have health insurance.

   (c) Jong earns \$450 per week plus $2\frac{1}{4}\%$ of the sale of goods on the first \$90 000 and $\frac{1}{2}\%$ on the remainder per week. Over the year, she averages sales of \$160 000 per week. She is single, has private health insurance and has 4 weeks unpaid holidays.

2. Stephanie works full time with an annual salary of \$78 000. She is single, without children and does not have private health insurance. Stephanie is considering working part-time.
   (a) Find the percentage change to her weekly take home pay using the tax scale above if she reduces her time fraction to 0.9, 0.8, 0.7, 0.6 of her current salary.

   (b) Comment on the type of reduction and find the equation of the line of best fit relating the percentage of salary P for time fraction t.

Consumer Arithmetic (2)

Interest is the fee charged when money is borrowed. The amount borrowed is called the principal, P and the interest rate is $r\%$ p.a. (per annum) for a period of time t.

Simple Interest

Simple interest, $S.I.$ is calculated at a percentage rate $r\%$ p.a. of the amount borrowed, P where the amount stays the same each year, t. The formula is

$$S.I. = P \times \frac{r}{100} \times t$$

1. Find the interest payable when:

   (a) \$4 000 is borrowed at 8% p.a. simple interest for 5 years

   (b) \$8 500 is borrowed at $6\frac{1}{2}\%$ p.a. simple interest for $2\frac{1}{3}$ years

   (c) \$3 500 is borrowed for $4\frac{3}{4}$ years at the rate of 8.2% p.a.

2. \$4 000 is borrowed to buy a television at the rate of 12% p.a. over a 3 year time period.

   Find:

   (a) the total interest payable

   (b) the total amount to be repaid

   (c) the amount to be repaid each month if equal instalments are to be made for the life of the loan.

3. When a total of \$1 200 simple interest is charged, find the value of the pronumeral when:

   (a) a principal amount, P is invested at $5\frac{1}{3}\%$ p.a. for 3 years

   (b) \$14 000 is invested at $r\%$ p.a. for 4 years

   (c) \$4 000 is invested at 4% for t years.

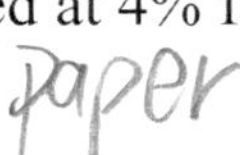

Consumer Arithmetic (2) page 2

4. Choose the better investment.
   Plan A: Investing an amount of money at 5% p.a. for 3 years then reinvesting the same amount for 3 years at 8%.
   Plan B: Investing the same amount as for Plan A at 6% for 7 years.

Compound Interest

Compound interest is calculated adding the interest to the principal before the next year's interest is calculated – it is a growing investment. Compound interest, $C.I.$ can be found by finding the interest each year, t adding it to the principal, P then finding the interest on the increased amount or by using the formula $C.I. = P((1+\frac{r}{100})^t - 1)$. The new principal P_{new} can be found using the formula $P_{new} = P(1+\frac{r}{100})^t$.

5. Find the value of an investment of \$20 000 that compounds at the rate of 4% p.a. for three years by:
   (a) using arithmetic to show the value at the end of each year

   (b) using the formula.

6. Find the simple interest rate that would deliver the same return for the same time period as an investment of \$3 200 compounding at the rate of 8% p.a. for 5 years.

7. Find the difference of the value and the percentage difference of an investment of \$5 000 that compounds at 12% p.a. yearly compared for the same investment compounding at 12% p.a. monthly, both invested for 4 years.

8. If the principal amount P is invested compounding at r% p.a. for t years, derive the formula for the value of the investment and hence, find the formula for the compound interest made.

Prime Numbers

1. Place the digits {0, 1, 2, 3, 4, 5, 6, 7} in the shaded squares so that the sum of adjacent numbers are all prime.

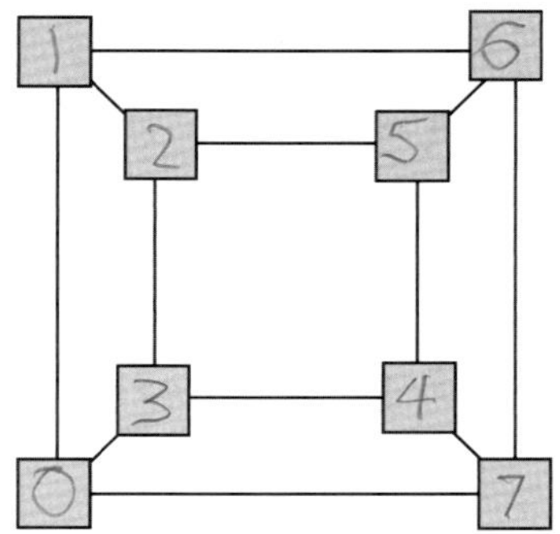

2. (a) List the first ten prime numbers after the number 3.

   5, 7, 11, 13, 17, 19, 23, 29, 31, 37

   (b) For the prime numbers listed in part (a), investigate the qualities of the values of the square of each number less one.

3. For prime numbers greater than 2:

   $3 = \left[\frac{1}{2}(3+1)\right]^2 - \left[\frac{1}{2}(3-1)\right]^2 = 4 - 1 = 3$, $5 = \left[\frac{1}{2}(5+1)\right]^2 - \left[\frac{1}{2}(5-1)\right]^2 = 9 - 4 = 5$, and so on, show that this calculation is true for the next 6 prime numbers in this sequence.

4. (a) List the prime numbers between 1 and 400.

   1, 3, 5, 7, 11, 13, 17, 19, 23, 29, 31, 37, 41, 43, 47, 49, 51, 53, 57, 59, 61, 67, 71, 73, 79, 83, 87, 89, 91, 93, 97, 101, 103,

   (b) Try to express the first 20 square numbers as the sum of two primes – not all are possible. 200□# + 2(1, 3, 5, 7, 11) ≠

Prime Numbers page 2

4. (c) Some numbers look as though they should be prime numbers. Explain why the following numbers are not prime numbers.

(i) 391 (ii) 377 (iii) 341

(iv) 301 (v) 221 (vi) 217

They can be multiplied by more that 1 and itself.

5. (a) Prove that for any two-digit number when it has its digits reversed, then the difference between the numbers can never be a prime number.
For example 21 – 12 = 9 and this is not a prime number.

Because that is how the man who made math wanted it to be.

(b) Prove that this also holds for:

(i) three-digit numbers

same

(ii) four-digit numbers.

same

Factors

1. A deficient number is a number with the sum of its factors, without the original number, is less than the number. The number 10 is a deficient number as the factors of 10, without the value of 10, are {1, 2, 5} and $1+2+5=8$, which is less than 10.
   (a) Show that 14 is a deficient number.

   (b) List the deficient numbers that are 30 or less.

2. An abundant number is one where the sum of its factors, without the original number, is greater than the number. 12 is an abundant number as the factors of 12, without the value of 12, are {1, 2, 3, 4, 6} and 1 + 2 + 3 + 4 + 6 = 16, which is greater than 12. Show that the following are abundant numbers.
   (a) 18 (b) 20 (c) 24

   (d) 30 (e) 36 (f) 40

   (g) 42 (h) 48

3. A perfect number is equal to the sum of its factors excluding the number itself. The first perfect number is 6 as 1 + 2 + 3 = 6.
   (a) Show that 28 is the next perfect number.

   (b) An algebraic definition of a perfect number is: $2^n(2^{n+1}-1)$, for $n \in Z^+$ for some values of n.
      (i) Show that when $n=1$ the value of $2^n(2^{n+1}-1) = 6$.

      (ii) Find the value of n so that $2^n(2^{n+1}-1) = 28$.

      (iii) Find the perfect number for $n=4$ and show it to be a perfect number.

      (iv) Use the expression from (b) to find the perfect numbers for $n = \{6, 12, 16, 18\}$.

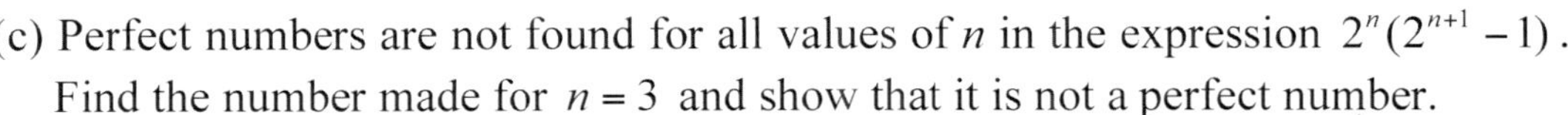

   (c) Perfect numbers are not found for all values of n in the expression $2^n(2^{n+1}-1)$. Find the number made for $n=3$ and show that it is not a perfect number.

Multiples and Factors

1. The numbers 56, 112, 24 and 216 have a number of factors in common. State the common factors by generating the prime factors for each number.

2. (a) State the prime factors of the following using index notation and hence, list the factors for each number.

   (i) 24 (ii) 196 (iii) 108 (iv) 200

   (b) Use the prime factors written in index form for the number 24 to find a connection between the powers of the prime factors and the number of factors it contains.

   (c) Use the results from (b) to state the number of factors of a number with the following prime number structures.

   (i) a^n (ii) $a^n \times b^m$ (ii) $a^n \times b^m \times c^k$

   (d) State the following numbers as a product of their prime factors and hence, state the number of factors for each number.

   (i) 9 (ii) 36

   (iii) 144 (iv) 10 125

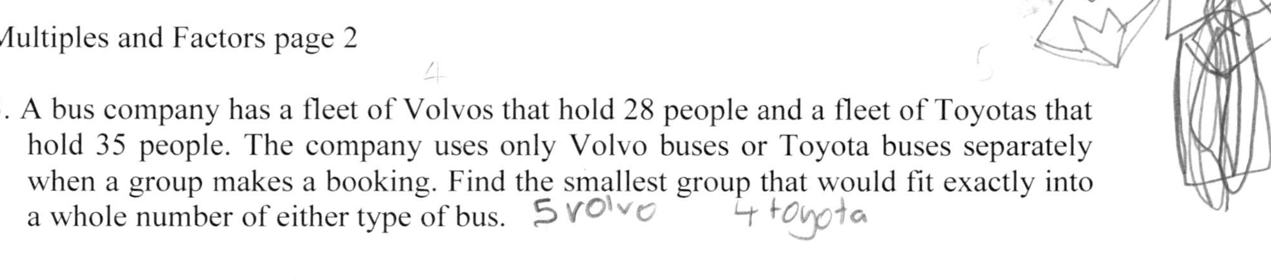

Multiples and Factors page 2

3. A bus company has a fleet of Volvos that hold 28 people and a fleet of Toyotas that hold 35 people. The company uses only Volvo buses or Toyota buses separately when a group makes a booking. Find the smallest group that would fit exactly into a whole number of either type of bus.

4. Without division, find the number of times that $900\,(900 = 2^2 \times 3^2 \times 5^2)$ will divide into the following numbers by first expressing them as a product of their prime factors.
(a) 64 800 (b) 135 000 (c) 4 374 000

5. A lolly distributor makes lollies in three colours, red, green and white. They want to sell them to shops in large bags that contain only one colour red, green or white and contain as many of the day's production as possible. Each day 1 225 000 red, 24 010 000 green and 17 150 000 white lollies are produced. How many lollies should go into each bag so that there are none left over?

Numbers and Algebra

1. An interesting algebraic relationship exists: $1 + 2 + 4 + 8 + \ldots\ldots + 2^k = 2^{k+1} - 1$.
   (a) Show that this holds for $k = 0,\ 1,\ 2,\ 3,\ 4,\ 5$.

   (b) Prove that the numbers generated must be odd for all values of k using both sides of the formula.

2. A relationship between the square of a number and the sum of cubes is $\left(\frac{1}{2}n(n+1)\right)^2 = 1^3 + 2^3 + 3^3 + \ldots. + n^3$. Show that this relationship holds for $n = \{1,\ 2,\ 3,\ 4,\ 5,\ 6\}$

3. The set of triangular numbers are based on the pattern of points placed in a triangular shape such as

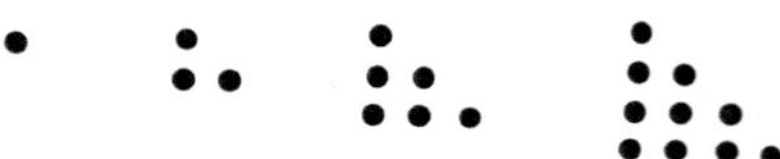

   (a) State the first ten triangular numbers.

   (b) Describe the way that the triangular numbers are generated.

   (c) Show that the sum of any pair of adjacent triangular numbers is a square number and make geometric sense of those results.

Numbers and Algebra page 2

3. (d) Show that the triangular numbers can be calculated using $\frac{1}{2}n(n+1)$, for $n = \{1, 2, 3, 4, 5, 6, 7, 8, 9, 10\}$.

4. For $n = \{0, 1, 2, 3, 4, 5\}$ show that $(n+1)^2 - n^2 = 2n+1$ and describe the numbers generated.

5. Fermat numbers are of the form $F_n = 2^{2^n} + 1$, for $n \in Z^+ \cup \{0\}$.

(a) Find the following Fermat numbers.

(i) F_0 (ii) F_1 (iii) F_2

(iv) F_3 (v) F_4

(b) Prove that all Fermat numbers are odd.

Numbers and Indices (1)

1. (a) Find the number of digits in the following.

   (i) 2^5 (ii) 2^8 (iii) 2^{10}

   (iv) Describe the relationship between the powers of 2 with the increase of the number of digits in the numbers.

   (b) Find the least value of x for which the following expressions have four digits.

   (i) 3^x (ii) 4^x (iii) 5^x

   (c) Explain the following equivalences.

   (i) $4^4 = 2^8 = 256$ (ii) $9^6 = 3^{12} = 531\ 441$

   (d) Find the values of x.

   (i) $2^x = 256^2$ (ii) $5^x = 125^4$

   (iii) $9^x = 3^{10}$ (iv) $16^x = 2^{12}$

2. There is a special relationship between the values 3, 4, and 5 so that $3^2 + 4^2 = 25 = 5^2$.

   (a) Find the next smallest set of three numbers that makes the relationship $a^2 + b^2 = c^2$ true.

   (b) Show that the first four multiples of the set of numbers {3, 4, 5} also satisfies the above relationship.

   (c) Find the missing values.

   (i) $7^2 + 24^2 = c^2$ (ii) $20^2 + 21^2 = c^2$ (iii) $a^2 + 12^2 = 13^2$

   (iv) $a^2 + 15^2 = 17^2$ (v) $21^2 + b^2 = 35^2$ (vi) $15^2 + b^2 = 39^2$

Numbers and Indices (2)

1. The rule for negative indices is $a^{-n} = \dfrac{1}{a^n}$.

   (a) Use the above rule to evaluate the following.

   (i) 4^{-1} (ii) 10×2^{-1}

   (iii) $16^{-1} \times 2^5$ (iv) $\dfrac{1}{4^{-1}}$

   (b) Evaluate the following for $x = 4$ and $y = 3$.

   (i) $x^2 + y^2$ (ii) $x^{-1} + y^{-1}$ (iii) $(x + y)^{-1}$

   (iv) $(2^{-x} + 2^{-y})^{-1}$ (v) xy^{-1} (vi) $\left(\dfrac{1}{x}y^{-1}\right)^{-1}$

2. For integers a and b, the operation is defined $a \otimes b = a^b + b^a$. Find the value of the following.

   (a) $a = 1,\ b = 2$ (b) $a = 1,\ b = 3$ (c) $a = 2,\ b = 3$ (d) $a = 2,\ b = 4$

   (e) $a = -1,\ b = 2$ (f) $a = 1,\ b = -2$ (g) $a = -1,\ b = -2$ (h) $a = -2,\ b = -3$

3. If $a^{2b} = 8$, find the value of the following.

   (a) $3a^{2b}$ (b) a^{4b} (c) $4a^{6b}$ (d) $3a^{-8b}$

4. Find the last digit in the following.

   (a) 2^5 (b) 2^{10} (c) 2^{20}

   (d) 2^{40} (e) 2^{99} (f) 2^{101}

5. Find all the values of a and b so that $a^b = b^a$.

Numbers and Square Roots

1. The first set of numbers for $\sqrt{a^2+b^2+c^2}=d$ where $a,\ b,\ c,\ d \in Z^+$ is $\sqrt{2^2+2^2+1^2}=\sqrt{9}=3$.

   (a) Find the numbers to make these relationships true:

   (i) $\sqrt{34^2+37^2+38^2}=d$ (ii) $\sqrt{19^2+22^2+26^2}=d$

   (iii) $\sqrt{4^2+b^2+7^2}=9$ (iv) $\sqrt{a^2+6^2+7^2}=11$

   (v) $\sqrt{a^2+3^2+6^2}=7$ (vi) $\sqrt{a^2+b^2+12^2}=17$

   For $n \in Z^+$ the values $n,\ n+1,\ n+2,...$ are consecutive numbers so that for $n=5$; $n,\ n+1,\ n+2,...=5,\ 5+1,\ 5+2,...=5,\ 6,\ 7,...$

   (b) Find the following for the values $n=1,\ 2$ and 3.

   (i) $\{n,\ n+1,\ n+2\}$ (ii) $\{n,\ n+2,\ n+4\}$

   (iii) $\{n,\ n,\ n+1\}$ (iv) $\{n,\ n+1,\ n+5\}$

   (c) Show that the following are true for the indicated n value.

   (i) $\sqrt{n^2+(n+1)^2+(n+1)^2}=41$; for $(n=23)$

   (ii) $\sqrt{n^2+(n+3)^2+(n+3)^2}=33$; for $(n=17)$

   (iii) $\sqrt{n^2+(n+1)^2+(n+3)^2+(n+5)^2}=15$; for $(n=5)$

   (iv) $\sqrt{n^2+(n+4)^2+(n+4)^2+(n+5)^2}=15$; for $(n=4)$

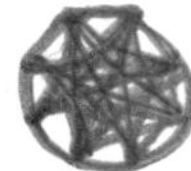

   (d) Find the value of n and hence, find the values to make the following true.

   (i) $\sqrt{n^2+(n+3)^2+16^2}=21$ (ii) $\sqrt{n^2+(n+3)^2+16^2}=25$

   (iii) $\sqrt{n^2+(n+1)^2+(n+2)^2+(n+6)^2}=21$ (iv) $\sqrt{n^2+(n+5)^2+(n+5)^2+(n+7)^2}=27$

   (v) $\sqrt{n^2+(n+4)^2+(n+6)^2+(n+7)^2}=29$ (vi) $\sqrt{n^2+(n+1)^2+(n+5)^2+(n+7)^2}=33$

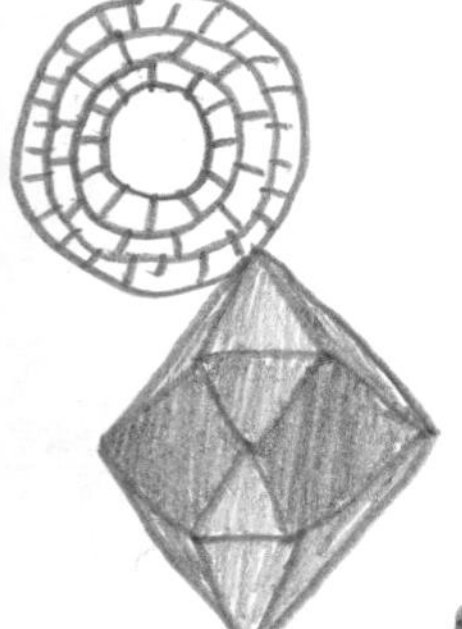

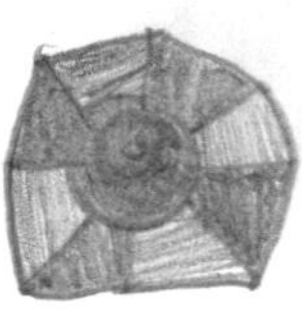

Number Patterns (1)

1. Calculate the values of the following for part (i) and hence, write the values of the expression in part (ii) without calculating.

   (a) (i) 4^2, 34^2, 334^2

   (ii) $3\ 333\ 334^2$, $33\ 333\ 334^2$, $333\ 333\ 334^2$, $3\ 333\ 333\ 334^2$

   (b) (i) 7^2, 67^2, 667^2

   (ii) $6\ 666\ 667^2$, $66\ 666\ 667^2$, $666\ 666\ 667^2$, $6\ 666\ 666\ 667^2$

   (c) (i) 9^2, 99^2, 999^2

   (ii) $9\ 999\ 999^2$, $99\ 999\ 999^2$, $999\ 999\ 999^2$, $9\ 999\ 999\ 999^2$

   (d) (i) 7×9, 77×99, 777×999

   (ii) $777\ 777 \times 999\ 999$, $7\ 777\ 777 \times 9\ 999\ 999$, $77\ 777\ 777 \times 99\ 999\ 999$

   (e) (i) $2\ 178 \times 4$, $21\ 978 \times 4$, $219\ 978 \times 4$

   (ii) $219\ 999\ 978 \times 4$, $2\ 199\ 999\ 978 \times 4$, $21\ 999\ 999\ 978 \times 4$

2. (a) Show that the number ***3 816 547 290*** has the quality that the left most n digits are exactly divisible by n and state another way that it is special.

   (b) Find the (i) smallest and (ii) largest ten-digit number that has the quality that the left most n digits are exactly divisible by n if digits can be used more than once.

3. (a) Evaluate the following and state the relationship that holds for each.

   (i) $(1+5+4+7)(1^2+5^2+4^2+7^2)$ (ii) $(2+1+9+6)(2^2+1^2+9^2+6^2)$

   (b) Show that the same relationship holds for:

   (i) 133 (ii) 315 (iii) 803 (iv) 1148

Number Patterns (1) page 2

4. When a number is cubed, it is multiplied by itself three times.
   $1^3 = 1 \times 1 \times 1 = 1,\ 2^3 = 2 \times 2 \times 2 = 8,\ 3^3 = 3 \times 3 \times 3 = 27$
   (a) Show that the first three cubes are made up of the sum of odd numbers – 1 odd number for the first cube, 2 odd numbers for the second, 3 for the third and so on.

   (b) Continue the pattern through for the first 9 cubes and state a general rule for this pattern.

5. (a) (i) Show that the sum of the first 5 odd numbers is 5^2.

   (ii) Find the sum of the first n odd numbers for $n \in \{1, 2, 3, ...7, 8, 9\}$, and show that each sum is equal to n^2.

   (iii) Let n stand for the first odd number, then find an expression for the sum of the first 5 odd numbers in terms of n and hence, show the sum to be equal to 5^2.

   (b) The sum of a number of terms in a sequence is given by the formula $S_n = \frac{n}{2}(2a + (n-1)d)$, where a is the first number, d is the difference between the numbers in the pattern and n is the number of terms. Using the formula show that the sum of the first n odd numbers is n^2.

Number Patterns (2)

1. The table shows the Fibonacci numbers across the top row and in the first column. It shows a summary of which Fibonacci numbers are factors of other Fibonacci numbers.

2 is a factor of 8

Place	3rd	4th	5th	6th	7th	8th	9th	10th	11th	12th	13th	14th	15th	16th
Number	2	3	5	8	13	21	34	55	89	144	233	377	610	987
F(3) – 2	✓	✗	✗	✓										
F(4) – 3						✓	✗	✗	✗					
F(5) – 5	✗	✗	✓											
F(6) – 8			✗	✓	✗	✗								
F(7) – 13				✗	✓									

8 is not a factor of 21

(a) Complete the table using ticks or crosses.

(b) Complete the following statements.

Every *3*rd Fibonacci number has a factor of 2, that is, has a factor of F(*3*).

Every *4*th Fibonacci number has a factor of___, that is, has a factor of F(__).

Every *5*th Fibonacci number has a factor of___, that is, has a factor of F(__).

Every *6*th Fibonacci number has a factor of___, that is, has a factor of F(__).

Every ___th Fibonacci number has a factor of___, that is, has a factor of F(__).

(c) Use the results above to make a general statement for the kth Fibonacci number.

2. The Fibonacci sequence is
$t_1 = 1,\ t_2 = 1,\ t_3 = 2,\ t_4 = 3, \ldots$ where $t_{n+2} = t_n + t_{n+1},\ n \in N$.

(a) Show that the first ten numbers in the sequence are $\{1, 1, 2, 3, 5, 8, 13, 21, 34, 55\}$, hence explain the sequence rule $t_{n+2} = t_n + t_{n+1}$.

(b) Find another relationship involving the difference between numbers in the sequence and express the sequence rule in terms of t_n, t_{n+1} and t_{n+2}.

Number Patterns (2) page 2

2. The 11th to the 40th values of the Fibonacci sequence are {89, 144, 233, 377, 610, 987, 1 597, 2 584, 4 181, 6 765, 10 946, 17 711, 28 657, 46 368, 75 025, 121 303, 196 418, 317 811, 514 229, 832 040, 1 346 269, 2 178 309, 3 524 578, 5 702 887, 9 227 465, 14 930 352, 24 157 817, 39 088 169, 63 245 986, 102 334 155}

   (c) A related sequence is defined as $T_n = \frac{t_{2n}}{t_n}$ such that $T_1 = \frac{t_2}{t_1} = \frac{1}{1} = 1$,

   $T_2 = \frac{t_4}{t_2} = \frac{3}{1} = 3$, $T_3 = \frac{t_6}{t_3} = \frac{8}{2} = 4$. Use the values of the Fibonacci sequence given above to find T_1, T_2, T_3,..., T_{19}, T_{20} and show this sequence to be Fibonacci-like.

3. Find the prime factors of the first twenty-five Fibonacci numbers and make a conclusion about the prime factors for those numbers.

Solutions

Cross-number Puzzle (1)

[1] 8	[2] 3		[3] 1	1	[4] 4		[5] 7	[6] 5
[7] 6	3	1	7		[8] 4	4	1	0
	6		[9] 1	[10] 7	2		3	
[11] 3	4	[12] 2		6		[13] 6	8	[14] 4
5		[15] 6	2		[16] 3	8		9
[17] 1	[18] 5	6		[19] 7		[20] 2	[21] 2	5
	3		[22] 2	2	[23] 8		0	
[24] 4	1	7	6		[25] 4	0	5	[26] 4
[27] 3	8		[28] 2	8	1		[29] 2	5

Cross-number Puzzle (2)

[1] 4	[2] 7	2	[3] 1		[4] 4	7	[5] 5	[6] 2
[7] 8	9		[8] 5	0	0		[9] 6	0
9		[10] 6	0		[11] 9	[12] 6		0
[13] 8	[14] 4	4	0		[15] 7	6	[16] 4	0
	3						0	
[17] 4	9	[18] 1	[19] 9		[20] 1	[21] 9	0	[22] 1
4		[23] 5	0		[24] 1	6		9
[25] 4	[26] 8		[27] 1	0	0		[28] 6	1
[29] 1	0	0	0		[30] 4	8	2	9

Cross-number Puzzle (3)

1.

[1] 1	[2] 9	6		[3] 3	5
[4] 4	2		[5] 2	8	
	[6] 6	2	7		[7] 1
[8] 2	1		[9] 4	9	0
9		[10] 9	4		2
[11] 4	4	1		[12] 4	9

2.

[1] 8	[2] 9	[3] 1		[4] 1	6
[5] 5	9	1		0	
8			[6] 1	0	[7] 9
		[8] 9	0		8
[9] 3	1	9		[10] 4	1
3		[11] 9	0	9	

Cross-number Puzzle (4)

1. (a)

	[1] 2	5	[2] 6	
[3] 2	5		[4] 4	[5] 9
[6] 1	6			6
6		[7] 3	[8] 6	1
		[9] 6	4	

1. (b)

	1 1	2	2 1	
3 3	6		4 6	5 4
6 4	9			8
3		7 1	8 4	4
	9 1	6	9	

2.

	1 4	2 8	3 4	
4 4	4	1		5 3 or 9
6 8	1		7 3	6
8 4		9 3	6	1
	10 3 or 9	6	1	

Arithmetic Challenge

1. (a) a = 8, b = 5, c = 1, d = 6, e = 2
   (b) a = 6, b = 8, c = 2, d = 4
   (c) a = 8, b = 0, c = 0, d = 2, e = 9
   (d) a = 8, b = 1, c = 5, d = 2, e = 5, f = 9
   (e) a = 9, b = 8, c = 7, d = 8, e = 8, f = g = 9, h = 6
   (f) a = 1, b = 1, c = 3, d = 8
   (g) a = 3, b = c = 0, with d = 9

2. (a) a = 3, b = 5
   (b) a = 7, b = 8
   (c) a = 3, b = 4, c = 9

3. (a) For a = 1, c = 1 or a = 0, c = 0; b = 6, d = e = f = 9

(b)

a	b	c	d	e
0	9	4	3	1
1	9	4	2	1
2	9	4	1	1
3	9	4	0	1

(c)

a	b	c
1	1	2
2	1	3
3	1	4
4	1	5
5	1	6
6	1	7
7	1	8
8	1	9

(d)

a	b	c
1	2	4
1	3	6
1	4	8

(e)

a	b	c	d
8	0	1	7
8	1	1	8
8	2	1	9
8	3	2	0
8	4	2	1
8	5	2	2
8	6	2	3
8	7	2	4
8	8	2	5
8	9	2	6

(f)

a	b	c	d	e	f
0	2	1	1	9	4
1	2	1	1	9	5
2	2	1	1	9	6
3	2	1	1	9	7
4	2	1	1	9	8
5	2	1	1	9	9
6	2	1	2	0	0
7	2	1	2	0	1
8	2	1	2	0	2
9	2	1	2	0	3
6	2	2	1	0	0
7	2	2	1	0	1
8	2	2	1	0	2
9	2	2	1	0	3

Fractions

1. (a) (i) $\frac{1}{3}$ (ii) $\frac{1}{4}$ (iii) $\frac{1}{5}$ (iv) $\frac{1}{n}$

   (b) (i) 2 (ii) $2\frac{1}{2}$ (iii) 3 (iv) $\frac{1}{2}(n+1)$

   (c) (i) $\frac{2}{3}$ (ii) $\frac{1}{2}$ (iii) $\frac{3}{5}$ (iv) $\frac{1}{2}$ (v) $\frac{n+1}{2n}$ (vi) $\frac{1}{2}$, for $n \geq 2$.

2. (a) $2\frac{5}{12}$ (b) $1\frac{1}{4}$ (c) $2\frac{12}{29}$ (d) $1\frac{1}{5}$

Directed Numbers (1)

1.

3: $2a+b+2c+d$ 2: $a+b+2c+d$ 1: $2a+2b+2c+d$ 0: $a+2b+2c+d$

−1: $2a+b+3c+2d$ −2: $a+3b+2c+d$ −3: $3a+b+c+d$ −4: $2a+b+c+d$

−5: $a+b+c+d$ −6: $2a+2b+c+d$

2. (a) (i) $ac+bd=12+3=15$ (ii) $ad+bc=-2-18=-20$

   (b) (i) $ab+d=2\times-3-1=-7$ (ii) $ad+c=2\times-1+6=4$ (iii) $bd+a=-3\times-1+2=5$

   (iv) $cd+a=6\times-1+2=-4$ (v) $ac+b=2\times6-3=9$ (vi) $cd+b=6\times-1-3=-9$

1. (c)

$$abc + d = (2 \times -3 \times 6) - 1 = -37$$
$$abd + c = (2 \times -3 \times -1) + 6 = 12$$
$$bcd + a = (-3 \times 6 \times -1) + 2 = 20$$
$$acd + b = (2 \times 6 \times -1) - 3 = -15$$

3.

$$a + b + c + d = -6 - 7 - 8 - 9 = -30$$
$$-a + b + c + d = 6 - 7 - 8 - 9 = -18$$
$$a - b + c + d = -6 + 7 - 8 - 9 = -16$$
$$a + b - c + d = -6 - 7 + 8 - 9 = -14$$
$$a + b + c - d = -6 - 7 - 8 + 9 = -12$$
$$-a - b + c + d = 6 + 7 - 8 - 9 = -4$$
$$-a + b - c + d = 6 - 7 + 8 - 9 = -2$$
$$-a + b + c - d = 6 - 7 - 8 + 9 = 0$$
$$a - b - c + d = -6 + 7 + 8 - 9 = 0$$
$$a - b + c - d = -6 + 7 - 8 + 9 = 2$$
$$a + b - c - d = -6 - 7 + 8 + 9 = 4$$
$$-a - b - c + d = 6 + 7 + 8 - 9 = 12$$
$$-a - b + c - d = 6 + 7 - 8 + 9 = 14$$
$$-a + b - c - d = 6 - 7 + 8 + 9 = 16$$
$$a - b - c - d = -6 + 7 + 8 + 9 = 18$$
$$-a - b - c - d = 6 + 7 + 8 + 9 = 30$$

Directed Numbers (2)

1. (a) Opposite faces add to −7.
   (b) (i) $\frac{1}{2}$ (ii) $\frac{1}{4}$ (iii) $\frac{1}{6}$ (iv) $\frac{3}{4}$
2. (a) (i) $\pm1, \pm5$
   (a) (ii) $\pm7, \pm8, \pm13$
   (b) (i) 15, −13, 12, ± 11, −10, ±9, −5, 4, −3, ±1
   (b) (ii) 19, 17, −15, 14, 12, 10, −8, 7, 6, 5, −4, −1
3. (a) (i) 14 (ii) 8
   (b) 10, 6, 4, 4, 2, 0, 0, -2, −4, −4, −8, −6, −10, −14
   (c) All results are even, 7 answers are positive 2 are zero and 7 are negative. All sequences of consecutive numbers contain {4, 2, 0, 0, −2, −4}.
4. (a) (i) −14 (ii) −8
   (b) −10, −6, −4, −4, −2, 0, 0, 2, 4, 4, 8, 6, 10, 14
   (c) All results are even, 7 answers are positive, 2 are zero and 7 are negative. All sequences of consecutive numbers contain {4, 2, 0, 0, −2, −4}.

Ratios (1)

1. (a) 20:3 (b) 1:2
2. (a) 34, 51, 85 (b) 102:170:255
3. (a) Bottle A: water 192mL and orange concentrate 48mL, Bottle B: water 160mL and orange concentrate 80mL
   (b) 11:4 (c) 30mL
4. (a) 125 (b) 140
5. (a) 8:6:5 (b) $8(\pi+2):2(3\pi+8):(5\pi+16)$

Ratios (2)

1. (a) $\frac{a^2+b^2}{(a+b)^2}$ (b) $\frac{\sqrt{a^2+b^2}}{a+b}$ (c) $a=b$
2. (a) 133 (b) 3 610
3. (a) 8:9 (b) 4:3
4. (a) (i) 6:1 (ii) 9:1 (b) (6, 9, 1) (12, 18, 2) (18, 27, 3)
5. (a) 2 red marbles need to be added to the bag
   (b) Two blue marbles need to be added to the bag
   (c) The number of blue marbles to be added to the bag is two less the number of red marbles.
   (d) The number of red marbles to be added is 4 and four divided by the number of red marbles – the 3 samples that are possible: for 1 red add 8 red, for 2 red add 6 red, for 4 red add 5 red.

Percentages (1)

1. 17
2. (a) 20 (b) $\frac{x}{9}$
3. 50
4. \$45
5. (a) $12\frac{1}{2}$L (b) $66\frac{2}{3}$%
6. (a) \$47 000 (b) \$32 531.20 (c) –4.8%

Percentages (2)

1. (a) (i) 1% decrease (ii) 4% decrease (iii) 16% decrease
   (b) (i) 30% (ii) 50% (iii) 60%
2. (a) increased by 21% (b) decreased by 19%
   (c) increased by 300% (d) decreased by 75%
3. (a) (i) unchanged (ii) increased by 8%
   (b) width is decreased by 20%
4. (a) increase by $16\frac{2}{3}$% (b) decrease by $8\frac{1}{3}$% (c) decrease by 25%
5. (a) (i) increase by 100% (ii) increase by 300%
   (b) (i) increase by 300% (ii) increase by 700%

Consumer Arithmetic (1)

1. (a) \$11 676.48 (b) \$13 645 (c) \$43 374
2. (a) 91.6%, 82.5%, 73.2%, 65.0%
   (b) Linear, $P=88.4t+11.74$

Consumer Arithmetic (2)

1. (a) \$1 600 (b) \$1 289.17 (c) \$1 363.25
2. (a) \$1 440 (b) 5 440 (c) \$151.11
3. (a) \$7500 (b) $3\frac{1}{3}\%$ (c) $7\frac{1}{2}$ years
4. Plan B
5. \$20 800, \$21 632, \$22 497.28
6. 9.4%
7. \$193.53, 3.9%
8. New Principal: t[th] year: $P(1+\frac{r}{100})^t$, Compound Interest: t[th] year: $P(1+\frac{r}{100})^t - P = P\left((1+\frac{r}{100})^t - 1\right)$

Prime Numbers

1.

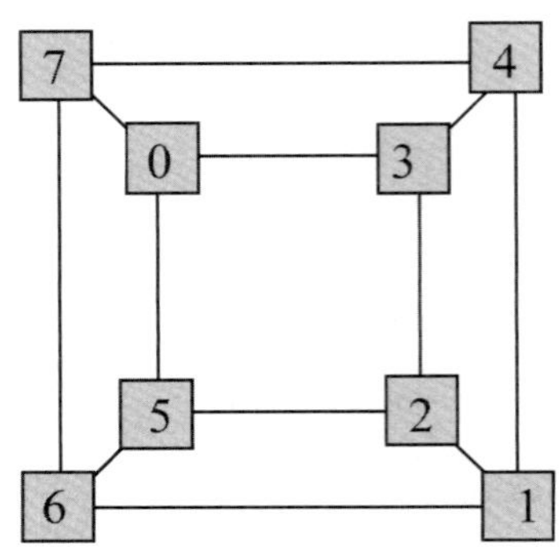

2. (a) 5, 7, 11, 13, 17, 19, 23, 29, 31, 37
   (b) The square of prime numbers greater than 3 less one are all divisible by 24.
4. (a) { 2, 3, 5, 7, 11, 13, 17, 19, 23, 29, 31, 37, 41, 43, 47, 53, 61, 67, 71, 73, 79, 83, 89, 97, 101, 103, 107, 113, 127, 131, 137, 139, 149, 151, 157, 163, 167, 173, 179, 181, 191, 193, 197, 199, 211, 223, 227, 229, 233, 239, 241, 251, 257, 263, 269, 271, 277, 281, 283, 293, 307, 311, 313, 317, 331, 337, 349, 353, 359, 367, 373, 383, 389, 397}
   (b) 1 impossible, 4 = 2 + 2, 9 = 2 + 7, 16 = 3 + 13, 11 + 5, 25 = 2 + 23, 36 = 13 + 23, 49 = 2 + 47, 64 = 17 + 47, 81 = 2 + 79, 100 = 3 + 97, 121 impossible, 144 = 47 + 97, 169 = 2 + 167, 196 = 29 + 167, 225 = 2 + 223, 256 = 89 + 167, 289 impossible, 324 = 101 + 223, 361 = 2 + 359, 400 = 17 + 383
   (c) The numbers have more than two factors.

Factors

1. (b) 2, 3, 4, 5, 7, 8, 9, 10, 11, 13, 14, 15, 16, 17, 19, 21, 22, 23, 25, 26, 27, 29, 30
3. (b) (ii) $n = 2$ (iii) 496 (iv) 8 128, 33 550 336, 8 589 869 056, 137 438 691 328
   (c) 120

Multiples and Factors

1. (a) The factors in common are: 2, 4 and 8.
2. (a) (i) $24 = 2^3 \times 3$, factors: {1, 2, 3, 4, 6, 8, 12, 24}
   (a) (ii) $196 = 2^2 \times 7^2$, factors: {1, 2, 4, 7, 98, 196}
   (a) (iii) $108 = 2^2 \times 3^3$ factors: {1, 2, 3, 4, 6, 9, 12, 18, 27, 36, 54, 108}

2. (a) (iv) $200 = 2^3 \times 5^2$ factors: {1, 2, 4, 5, 8, 10, 20, 25, 40, 50, 100, 200}
   (b) In general, the number of factors can be found by adding 1 to the power of each prime factor and multiplying those increased values.
   (c) (i) $n+1$ (ii) $(n+1)(m+1)$ (iii) $(n+1)(m+1)(k+1)$
   (d) (i) 3 factors (ii) 9 factors (iii) 15 factors (iv) 20 factors
3. 5 Volvo buses or 4 Toyota buses
4. (a) 72 times (b) 150 times (c) 4 860 times
5. 245 000

Numbers and Algebra

3. (b) The nth triangular number is the sum of the first n natural numbers.
4. The numbers generated is the set of odd numbers.
5. (a) (i) 3 (ii) 5 (iii) 17 (iv) 257 (v) 65 537
   (b) Each value must be odd

Numbers and Indices (1)

1. (a) (i) 2 digits (ii) 3 digits (iii) 4 digits
   (a) (iv) multiplying each value by $2^3 \approx 10$ increases it by an extra digit.
   (b) (i) 7 (ii) 5 (iii) 5
   (c) (i) 2^8 (ii) 3^{12}
   (d) (i) 16 (ii) 12 (iii) 5 (iv) 3
2. (a) 5, 12, 13
   (c) (i) 7 (ii) 29 (iii) 5 (iv) 8 (v) 28 (vi) 36

Numbers and Indices (2)

1. (a) (i) $\frac{1}{4}$ (ii) 5 (iii) 2 (iv) 4
   (b) (i) 25 (ii) $\frac{7}{12}$ (iii) $\frac{1}{7}$ (iv) $5\frac{1}{3}$ (v) $1\frac{1}{3}$ (vi) 12
2. (a) 12 (b) 4 (c) 17 (d) 32 (e) $1\frac{1}{2}$ (f) −1
   (g) $\frac{1}{2}$ (h) $-\frac{1}{72}$
3. (a) 24 (b) 64 (c) 2 048 (d) $\frac{3}{4096}$
4. (a) 2 (b) 4 (c) 6 (d) 6 (e) 8 (f) 2
5. For all a = b; a = 2, b = 4 and a = 4, b = 2

Numbers and Square Roots

1. (a) (i) 63 (ii) 39 (iii) 4 (iv) 6 (v) 2 (vi) a = 8, b = 9
   (b)

	$n=1$	$n=2$	$n=3$
(i)	1,2,3	2,3,4	3,4,5
(ii)	1,3,5	2,4,6	3,5,7
(iii)	1,1,2	2,2,3	3,3,4
(iv)	1,2,6	2,3,7	3,4,8

   (d) (i) 21 (ii) 25 (iii) 21 (iv) 27 (v) 29 (vi) 33

Number Patterns (1)

1. (a) (i) 16, 1 156, 111 556 (ii) 11 111 115 555 556, 1 111 111 155 555 556, 111 111 111 555 555 556, 11 111 111 115 555 555 556
   (b) (i) 49, 4 489, 444 889 (ii) 44 444 448 888 889, 4 444 444 488 888 889, 444 444 444 888 888 889, 44 444 444 448 888 888 889
   (c) (i) 81, 9 801, 998 001 (ii) 99 999 980 000 001, 9 999 999 800 000 001, 9 999 999 998 000 000 001, 99 999 999 980 000 000 001
   (d) (i) 63, 7 623, 776 223 (ii) 777 776 222 223, 77 777 762 222 223, 7 777 777 622 222 223
   (e) (i) 8 712, 87 912, 879 912 (ii) 879 999 912, 8 799 999 912, 87 999 999 912
2. (b) (i) 1 232 042 400 (ii) 9 876 545 640
3. (a) (i) 1 547 (ii) 2 196
4. (b) To find the nth cube add n consecutive odd numbers starting with the ath odd number where $a = \frac{1}{2}n^2 - \frac{1}{2}n + 1$ - note that the quadratic was found using technology.
5. (a) (i) 1, 4, 9, 16, 25, 36, 49, 64, 81 (ii) $5n + 20$

Number Patterns (2)

1. (a)

Place	3rd	4th	5th	6th	7th	8th	9th	10th	11th	12th	13th	14th	15th	16th
Number	2	3	5	8	13	21	34	55	89	144	233	377	610	987
F(3) - 2	✓	✗	✗	✓	✗	✗	✓	✗	✗	✓	✗	✗	✓	✗
F(4) - 3	✗	✓	✗	✗	✗	✓	✗	✗	✗	✓	✗	✗	✗	✓
F(5) - 5	✗	✗	✓	✗	✗	✗	✗	✓	✗	✗	✗	✗	✓	✗
F(6) - 8	✗	✗	✗	✓	✗	✗	✗	✗	✗	✓	✗	✗	✗	✗
F(7) - 13	✗	✗	✗	✗	✓	✗	✗	✗	✗	✗	✗	✓	✗	✗

   (b) 3, 5, 8, 13
   (c) Every kth Fibonacci number has a factor of F(k).
2. (b) $t_{n+2} - t_{n+1} = t_n$, $t_{n+1} = t_{n+2} - t_n$
   (c) {11, 18, 29, 47, 76, 123, 199, 322, 521, 843, 1 364, 2 207, 3 571, 5 778, 9 349, 15 127}
3. The Fibonacci numbers greater than 1 except the sixth (8) and twelfth (144) have at least one prime factor that is not a factor of any earlier Fibonacci number.

Chapter 2

General Algebra

Areas of Interest

Basic Algebra: Add and Subtract

Basic Algebra: Multiply and Divide

Algebraic Processes

Substitution and Number Machines

Pythagorean Triples

Numbers and Algebra

Basic Algebra: Add and Subtract

1. Complete the algebra grids using the clues.

(a)

1		2		3		4
5					6	
				7		
	8		9			
10			11			
12					13	

Across

1. $2a + 3b - a - 2b$ **3.** $2a - 3 - a$

5. $ac - 6ab + ac + ab$ **7.** $3b + c - 2b - 2c$

8. $14a + 5 - (a + 7)$ **10.** $2a - 3 - (2a - 1$

11. $3cdb - 2dcb$ **12.** $-(d - 4a)$

13. $5fe - 4ef$

Down

1. $2a + 2b - a - 4b$ **2.** $b - c - 3 + 2c$

3. $-c - 2b + 2a - 3b - a$

4. $2bc + 3 - cb$

6. $4de + a - 6ed$ **8.** $a + c - (c - 11a)$

9. $abd - adb + dba$ **10.** $3a - (3a - 14)$

(b)

1			2	3		4
			5			
8	9	10				
11					12	
13	14			15		

Across

1. $2a - 3b + a + b$ **5.** $-(3a - 2b) + 5a -$

8. $3dc - 2 - 2cd - 1$ **11.** $4 + a - (4 - a)$

12. $2 + a - (a - 10)$ **13.** $-(24b - c)$

16. $4d - c + b + d + 5c + 2b$

Down

1. $2b + 4c - (-b + 6c)$ **2.** $6 + bc + cb - 2$

3. $3b - (2b - d)$ **4.** $-(2c - 5a)$

9. $2a - c - (c - a)$ **15.** $dbc - cbd + bc$

Basic Algebra: Add and Subtract page 2

2. Add the expressions from the first column to the expressions in the first row.

(a)

+	**4**	**$-2a$**		**$-2a + 3b$**
7	11			
$-3a$			$b - 4a$	
				$4b - a$
		$3b - 3a$		

(b)

+	**8**		**$-a - 2b$**	
	$7a + 8$			
$-a-3$		$-1\frac{1}{3}a - 3$		
			$-3a - b$	
$4a-3b$				$9a - 6b$

3. Take the expressions from the first column from the expressions in the first row.

(a)

−	**9**	**$2a$**		**$a + 2b$**
3	6			
		$-a - 2$		
$a-b$			$3a + b$	
				$3a + b$

(b)

−	**$2a$**		**$5a - b$**	
	$a + 3$			
$2a$		$a - 2$		
			$6a - 3b$	
$-a-b$				$2b$

Basic Algebra: Multiply and Divide

1. Complete the algebra grids using the clues.

(a)

1		2		3		4
		5				
				6		
		7	8			
	9					
10			11			
12						

Across

1. $2a \times 4b$

3. $-4 \times -3a$

5. $-\frac{1}{2}(4-2d)$

6. $\frac{1}{2}(2a-2c)$

7. $\frac{1}{2}(4-2b)$

9. $7(bc+2)$

10. $2(a-1)+2$

11. $-(a+4)+a+1$

12. $2(b-2a+\frac{1}{2}c)$

Down

1. $2(4b+1)$

2. $b(d+2)$

3. $3a(4b+1)$

4. $bc(a+4)$

8. $-(c+4)$

9. $\frac{2}{3} \times 10\frac{1}{2}ab$

10. $\frac{11}{a} \times 2a$

(b)

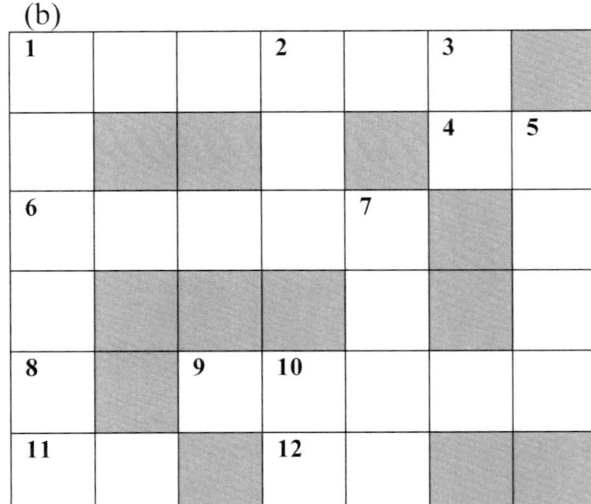

Across

1. $(x+1)(x+2)$

4. $-2y \times -2y$

6. $5(x+3)$

9. $-11(2x+6)+24x$

11. $\frac{11x}{3} \times \frac{18}{x}$

12. $4x-12(\frac{1}{2}+\frac{1}{3}x)$

13. $(x+2)(x+3)$

Down

1. $(x+3)(x+2)$

2. $\frac{1}{2}(2x-2)$

3. $\frac{12x}{7} \times \frac{14}{x}$

5. $(y-4)(y+4)$

7. $6(9-x)$

10. $\frac{1}{2}(2x-10)$

Basic Algebra: Multiply and Divide page 2

2. Multiply the expressions in the first column to the expressions in the first row.

(a)

$\times$	**4**	**$2a$**		
7	28	14a		$-28c$
$3a$				
$-3a$				
$-b$			$2b^2$	

(b)

$\times$		**-4**		
2a	$4a^2$			
$a+b$				
		$-4a+4b$	$ab-b^2$	
$2a+b$				$6a^2+3ab$

3. Show that the following expressions are equivalent using the values indicated.

(a) $3(2a+b)=6a+3b$, for:

(i) $a=2,\ b=3$ (ii) $a=-4,\ b=-5$ (iii) $a=\frac{1}{2},\ b=-1$

(b) $(a+b)^2=a^2+2ab+b^2$, for:

(i) $a=1,\ b=2$ (ii) $a=-1,\ b=4$ (iii) $a=\frac{1}{2},\ b=2$

Algebraic Processes

1. Multiply the following expressions by 4 and express as the sum of terms.

(a) $\frac{3a-2}{18a}$ (b) $\frac{4-2b}{2-4b}$ (c) $-\frac{3a+16}{4}$

(d) $\frac{5}{\frac{1}{2a}}-\frac{1\frac{1}{3}}{2a}$ (e) $\frac{5a+1}{\frac{1}{2}}$ (f) $\frac{2}{8a}-\frac{5a}{\frac{1}{4}}$

2. Multiply the following expressions by $-5a$ and express as the sum of terms.

(a) $\frac{a-2}{5}$ (b) $\frac{b}{5}+1$ (c) $\frac{a-1}{-15a}$

(d) $\frac{2a-3}{5}$ (e) $5-\frac{3a}{5}$ (f) $\frac{4a+7}{5}+1$

Algebraic Processes page 2

3. Divide the following expressions by $3a$ and express as the sum of terms.

(a) $\frac{1}{2}a$ (b) $-15a$ (c) $21a-36$

(d) $7a^2-12a$ (e) $2\frac{2}{5}a^2$ (f) $\frac{5}{9}a^2-15a+2\frac{1}{7}$

4. Add $-\frac{1}{4}(x+4)$ to the following expressions and express as the sum of terms.

(a) $-3(x+1)$ (b) $-2(x-3)$ (c) $3(2x-4)+12$

(d) $\dfrac{(6x+1)}{3}$ (e) $\frac{2}{5}(3-x)$ (f) $\frac{2}{3}(9x-12)$

5. Subtract $(\frac{1}{3}x-4\frac{3}{5})$ from the following expressions and express as the sum of terms in fractional form.

(a) $11\frac{1}{2}x-10$ (b) $\dfrac{x-6}{4}$ (c) $2\frac{1}{4}(-x+2)$

(d) $\dfrac{-6\frac{2}{3}+2x}{4}$ (e) $-\frac{2}{3}x-1.2$ (f) $\dfrac{\frac{2}{3}(-5x+11)}{-1}$

Substitution and Number Machines

1. Pass the x-values shown in the table through the number machines and work out the y-values that are produced. State the rule for each machine.

(a) (i)

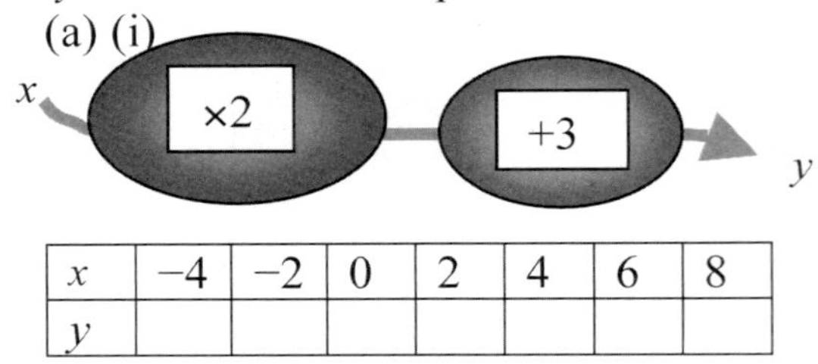

x	−4	−2	0	2	4	6	8
y							

(ii) Rule: y = ________________

(b) (i)

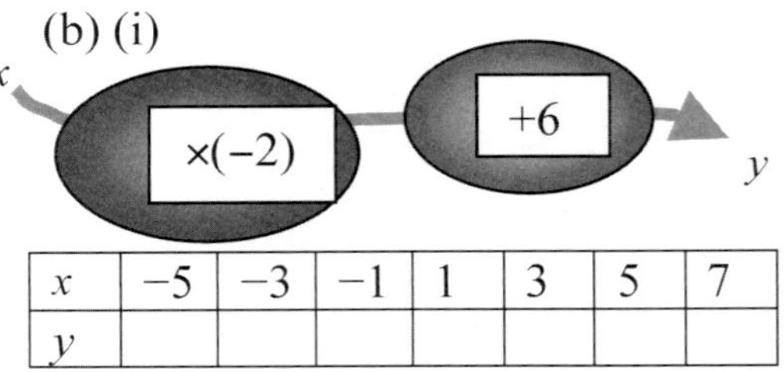

x	−5	−3	−1	1	3	5	7
y							

(ii) Rule: y = ________________

(c) (i)

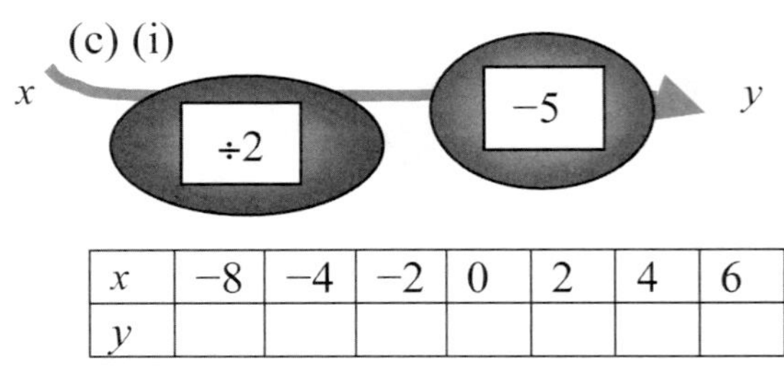

x	−8	−4	−2	0	2	4	6
y							

(ii) Rule: y = ________________

(d) (i)

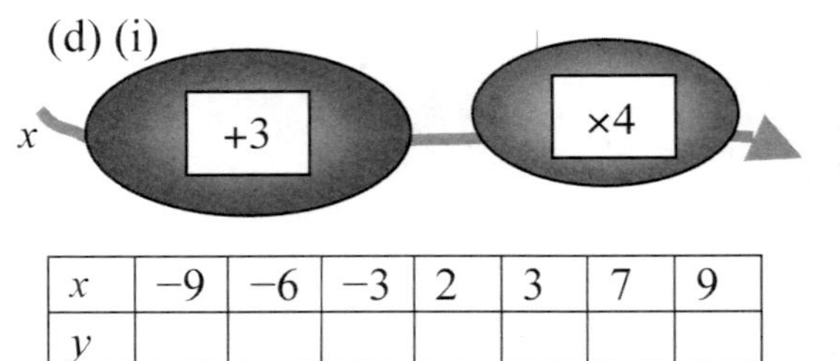

x	−9	−6	−3	2	3	7	9
y							

(ii) Rule: y = ________________

(e) (i)

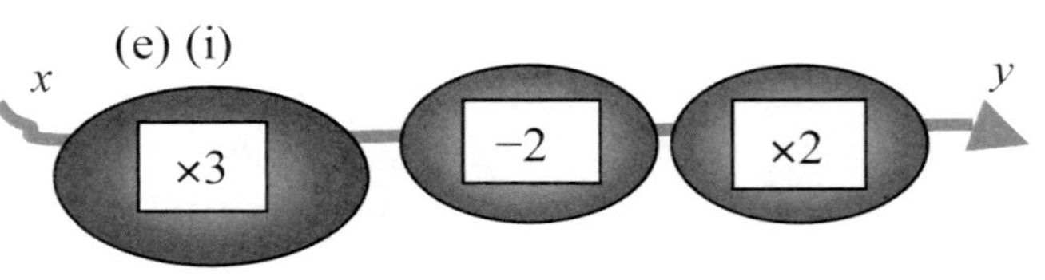

x	−10	−6	−1	0	1	2	3	5	8
y									

(ii) Rule: y = ________________

(f) (i)

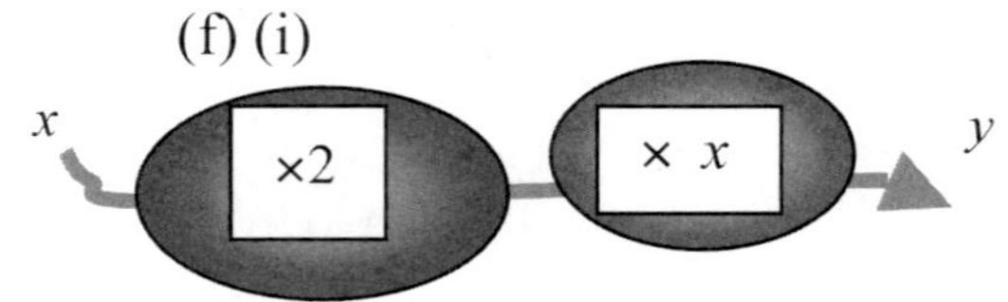

x	−8	−5	0	1	2	4	7
y							

(ii) Rule: y = ________________

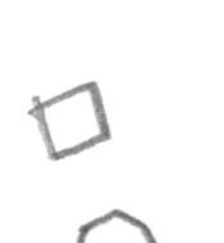

Substitution and Number Machines page 2

2. Complete the table of values for each equation.

(a) $y = 3x - 2$

x	−3	−1	0	2	5
y					

(b) $y = -4 - 5x$

x	−5	−2	−1	1	7
y					

(c) $y = 1 - \frac{1}{2}x$

x	−7	−3	−1	0	3
y					

3. Find the rule that was used to make each of the following table of values.

(a)

x	−2	−1	0	1	2
y	−3	−1	1	3	5

(b)

x	−3	−1	1	3	5
y	−23	−9	5	19	33

(c)

x	−4	−2	0	2	4
y	0	1	2	3	4

(d)

x	1	2	3	4	5
y	1	0	−1	−2	−3

(e)

x	−4	−2	0	2	4
y	3	2	1	0	−1

(f)

x	−5	−4	−3	−2	−1
y	15	13	11	9	7

Pythagorean Triples

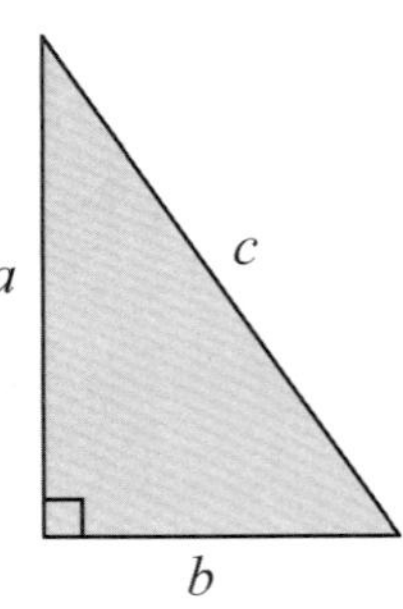

Pythagoras of the Greek Island of Samos popularised the relationship between the side lengths of a right-angled triangle, known as the theorem $a^2 + b^2 = c^2$, where c is the length of the hypotenuse and a and b are the other two side lengths.

1. Find the missing values.

(a) 6, 8, c (b) 5, 12, c (c) 12, 16, c (d) 27, 36, c (e) 13, 84, c

(f) a, 15, 17 (g) a, 21, 29 (h) a, 70, 74 (i) a, 48, 50 (j) a, 60, 61

(k) 7, b, 25 (l) 16, b, 34 (m) 40, b, 58 (n) 28, b, 53 (o) 12, b, 37

2. (a) Substitute the values given in the table to find values for $n^2 + 1$, $n^2 - 1$ and $2n$ and show each to be a Pythagorean triple.

n	2	3	4	5	6	7	8
$n^2 + 1$							
$n^2 - 1$							
$2n$							

(b) Use algebra to show that $(n^2 + 1)^2 = (n^2 - 1)^2 + (2n)^2$.

Pythagorean Triples page 2

3. (a) Substitute the values given in the table to find values for n^2+4, n^2-4 and $4n$ and show each to be a Pythagorean triple.

n	3	4	5	6	7	8
n^2+4						
n^2-4						
$4n$						

(b) Use algebra to show that n^2+4, n^2-4 and $4n$ are side lengths of a right-angled triangle.

4. A right-angled triangle is shown with side lengths m^2+n^2, m^2-n^2 and $4mn$ for $m > n$.

(a) Use the values for m and n to show that each of the following are Pythagorean triples.

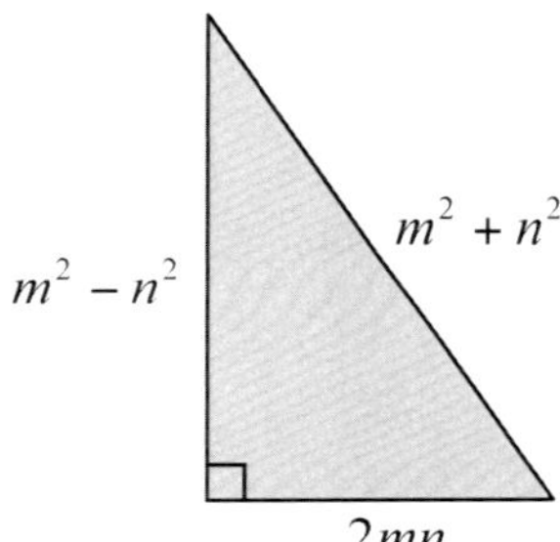

m	n	m^2+n^2	$2mn$	m^2-n^2
2	1			
3	1			
3	2			
4	1			
4	3			
5	4			

(b) Use algebra to show that the triangle is right-angled.

Numbers and Algebra

1. Some two-digit numbers are special in that the sum of the digits plus the product of the digits equals the original number.
   For example, 25 is not one of these numbers because, $2+5+2\times 5=17$, and $17\neq 25$.
   (a) Use algebra to show that any two-digit number ending with a particular digit is special according to the above process and state that digit.

   (b) List all the numbers that are special in this way and show that they satisfy the above condition.

2. (a) (i) Use the number 271 to rearrange the digits to make five different numbers.

   (ii) For each rearranged number and 271, subtract the larger number from the smaller number and show that the difference is always divisible by 9.

   (b) For the three-digit number *abc*, use algebra to show that the difference between the number *abc* and a different number, where the digits are arranged in a different order, is always divisible by 9.

Numbers and Algebra page 2

(c) (i) Use the number 228 to rearrange the digits to make two different numbers and repeat the same subtraction process from (a) to show that the difference between the numbers is divisible by 9.

(ii) Use algebra to show that the difference of all numbers of this type will always be divisible by 9.

3. (a) (i) Show that when the number 21 has its digits reversed the difference between the new number and 21 is 9.

(ii) Use algebra to find the connection between the digits of a two-digit number so that when they are reversed the difference between the numbers is 9, hence find the set of two digit numbers.

(b) Use algebra to find the connection between the digits of a two-digit number so that when they are reversed the difference between the numbers is 18, hence find the set of two-digit numbers.

(c) Use the results from above to make a conclusion about the difference between two-digit numbers when their digits are reversed.

Numbers and Algebra page 3

4. (a) (i) Make a two-digit number without a zero and reverse the digits to make a second number. Add the numbers together and divide the result by the sum of the digits. Repeat this process three more times choosing different sets of two-digit numbers. Comment on the results.

(ii) Try the process to see whether it works if one of the digits is zero.

4. (b) (i) Use the pronumerals a and b to prove the result observed in both sections of part (a).

(ii) Find a set of digits for which this process does not work.

(c) (i) Repeat the process for the two-digit numbers if one of the digits is negative. Use algebra to prove your observations and find values for the digits where the process does not work in each case.

(ii) Repeat the process for the two-digit numbers if both digits are negative. Use algebra to prove your observations and find values for the digits where the process does not work in each case.

Solutions

Basic Algebra: Add and Subtract

1.

(a)

1 a	+	**2** b		**3** a	−	4 3
−		+		−		−
5 2	a	c	−	5	**6** a	b
b		−		**7** b	−	c
	8 1	3	**9** a	−	2	
10 1	2		**11** b	c	d	
12 4	a	−	d		**13** e	f

(b)

1 3	a	−	**2** 2	**3** b		**4** 5
b			**5** b	+	2	a
8 −	**9** 3	**10** +	c	d		−
11 2	a		+		**12** 1	2
13 c	**14** −	2	4	**15** b		c
	2			c		
16 4	c	+	5	d	+	b

2. (a)

+	**4**	**−2*a***	**−*a* + *b***	**−2*a* + 3*b***
7	11	$7-2a$	$7-a+b$	$7-2a+3b$
−3*a*	$4-3a$	$-5a$	$b-4a$	$3b-5a$
a*+*b	$a+b+4$	$b-a$	$2b$	$4b-a$
−*a*+3*b*	$4-a+3b$	$3b-3a$	$4b-2a$	$6b-3a$

2. (b)

+	**8**	$-\frac{1}{3}a$	$-a-2b$	$5a-3b$
7a	$7a+8$	$6\frac{2}{3}a$	$6a-2b$	$12a-3b$
−a−3	$5-a$	$-1\frac{1}{3}a-3$	$-2a-2b-3$	$4a-3b-3$
−2a+b	$8-2a+b$	$-2\frac{1}{3}a+b$	$-3a-b$	$3a-2b$
4a−3b	$8+4a-3b$	$3\frac{2}{3}a-3b$	$3a-5b$	$9a-6b$

3. (a)

−	**9**	**2a**	**4a**	**a + 2b**
3	6	$2a-3$	$4a-3$	$a+2b-3$
3a+2	$7-3a$	$-a-2$	$a-2$	$-2a+2b-2$
a−b	$9-a+b$	$a+b$	$3a+b$	$3b$
−2a+b	$9+2a-b$	$4a-b$	$6a-b$	$3a+b$

(b)

−	**2a**	**3a−2**	**5a−b**	**−a + b**
a −3	$a+3$	$2a+1$	$4a-b+3$	$-2a+b+3$
2a	0	$a-2$	$3a-b$	$-3a+b$
−a+2b	$3a-2b$	$4a-2b-2$	$6a-3b$	$-b$
−a−b	$3a+b$	$4a+b-2$	$6a$	$2b$

Basic Algebra: Multiply and Divide

1.

(a)

1 8	*a*	**2** *b*	■	**3** 1	2	**4** *a*
b	■	**5** *d*	−	2	■	*b*
+	■	+	■	**6** *a*	−	*c*
2	■	**7** 2	**8** −	*b*	■	+
■	**9** 7	*b*	*c*	+	1	4
10 2	*a*	■	**11** −	3	■	*b*
12 **2**	*b*	−	4	*a*	+	*c*

1. (b)

1 x^2	+	3	2 x	+	3 2	
+			−		4 4	5 y^2
6 5	x	+	1	7 5		−
x				4		1
8 +		9 2	10 x	−	6	6
11 6	6		12 −	6		
	13 x^2	+	5	x	+	6

2.

(a)

×	**4**	**2*a***	**−2*b***	**−4*c***
7	28	$14a$	$-14b$	$-28c$
3*a*	$12a$	$6a^2$	$-6ab$	−12ac
−3*a*	$-12a$	$-6a^2$	$6ab$	12ac
−*b*	$-4b$	$-2ab$	$2b^2$	4bc

(b)

×	**2*a***	**−4**	***b***	**3*a***
2*a*	$4a^2$	$-8a$	$2ab$	$6a^2$
(*a*+*b*)	$2a^2+2ab$	$-4a-4b$	$ab+b^2$	$3a^2+3ab$
(*a*−*b*)	$2a^2-2ab$	$-4a+4b$	$ab-b^2$	$3a^2-3ab$
(2*a*+*b*)	$4a^2+2ab$	$-8a-4b$	$2ab+b^2$	$6a^2+3ab$

Algebraic Processes

1. (a) $\frac{2}{3}-\frac{4}{9a}$ (b) $\frac{8}{1-2b}-\frac{2b}{1-2b}$ (c) $-3a-16$
   (d) $40a-\frac{8}{3a}$ (e) $40a+4$ (f) $\frac{1}{a}-80a$
2. (a) $-a^2+2a$ (b) $-ab-5a$ (c) $\frac{a}{3}-\frac{1}{3}$
   (d) $-2a^2+3a$ (e) $-25a+3a^2$ (f) $-4a^2-12a$
3. (a) $\frac{1}{6}$ (b) -5 (c) $7-\frac{12}{a}$
   (d) $\frac{7a}{3}-4$ (e) $\frac{4a}{5}$ (f) $\frac{5a}{27}-5+\frac{5}{7a}$
4. (a) $-3\frac{1}{4}x-4$ (b) $-2\frac{1}{4}x-5$ (c) $5\frac{3}{4}x-1$
   (d) $1\frac{3}{4}x-\frac{2}{3}$ (e) $\frac{1}{5}-\frac{13}{20}x$ (f) $5\frac{3}{4}x-9$
5. (a) $11\frac{1}{6}x-5\frac{2}{5}$ (b) $-\frac{1}{12}x+3\frac{1}{10}$ (c) $-2\frac{7}{12}x+9\frac{1}{10}$
   (d) $-\frac{1}{6}x+2\frac{14}{15}$ (e) $-x+3\frac{2}{5}$ (f) $3x-2\frac{11}{15}$

Substitution and Number Machines

1. (a) (i) −5, −1, 3, 7, 11, 15, 19 (ii) $y=2x+3$
   (b) (i) 16, 12, 8, 4, 0, −4, −8 (ii) $y=-2x+6$
   (c) (i) −9, −7, −6, −5,−4, −3, −2 (ii) $y=\frac{1}{2}x-5$

(d) (i) $-24, -12, 0, 20, 24, 40, 48$ (ii) $y = 4(x+3)$

(e) (i) $-64, -40, -10, -4, 2, 8, 14, 26, 44$ (ii) $y = 2(3x-2)$

(f) (i) $128, 50, 0, 2, 8, 32, 98$ (ii) $y = 2x^2$

2. (a) $-11, -5, -2, 4, 13$
   (b) $21, 6, 1, -9, -39$
   (c) $4\frac{1}{2}, 2\frac{1}{2}, 1\frac{1}{2}, 1, -\frac{1}{2}$

3. (a) $y = 2x+1$ (b) $y = 7x-2$ (c) $y = \frac{1}{2}x+2$
   (d) $y = 2-x$ (e) $y = -\frac{1}{2}x+1$ (f) $y = -2x+5$

Pythagorean Triples

1. (a) 10 (b) 13 (c) 20 (d) 45 (e) 85 (f) 8
   (g) 20 (h) 24 (i) 14 (j) 11 (k) 24 (l) 30
   (m) 42 (n) 45 (o) 35

2. (a)

n	2	3	4	5	6	7	8
n^2+1	$4+1=5$	$9+1=10$	$16+1=17$	$25+1=26$	$36+1=37$	$49+1=50$	$64+1=65$
n^2-1	$4-1=3$	$9-1=8$	$16-1=15$	$25-1=24$	$36-1=35$	$49-1=48$	$64-1=63$
$2n$	4	6	8	10	12	14	16

3. (a)

n	3	4	5	6	7	8
n^2+4	$9+4=13$	$16+4=20$	$25+4=29$	$36+4=40$	$49+4=53$	$64+4=68$
n^2-4	$9-4=5$	$16-4=12$	$25-4=21$	$36-4=32$	$49-4=45$	$64-4=60$
$4n$	12	16	20	24	28	32

4. (a)

m	n	m^2+n^2	$2mn$	m^2-n^2
2	1	$4+1=5$	$2\times2\times1=4$	$2^2-1^2=3$
3	1	$3^2+1^2=10$	$2\times3\times1=6$	$3^2-1^2=8$
3	2	$3^2+2^2=13$	$2\times3\times2=12$	$3^2-2^2=5$
4	1	$4^2+1^2=17$	$2\times4\times1=8$	$4^2-1^2=15$
4	3	$4^2+3^2=25$	$2\times4\times3=24$	$4^2-3^2=7$
5	4	$5^2+4^2=41$	$2\times5\times4=40$	$5^2-4^2=9$

Numbers and Algebra

1. (a) The process holds for two-digit numbers ending with 9.
   (b) 19, 29, 39, 49, 59, 69, 79, 89, 99

2. (a) (i) For the number 271, the other rearranged numbers are 217, 127, 172, 721, 712.
   (c) (i) 282 and 822

3. (a) The set of numbers are such that the first digit is one more than the second digit: {21, 32, 43, 54, 65, 76, 87, 98}.
   (b) The set of numbers are such that the first digit is two more than the second digit: {31, 42, 53, 64, 75, 86, 97}.
   (c) When the digits of a two-digit number are reversed, then the difference between it and the original number is a multiple of 9. The greater the difference the greater the multiple of 9 that results.

4. (a) (i) The result is always 11 irrespective of the digits.
   (ii) It works when a digit is zero as long as the nature of the number is qualified.
   (b) (ii) $a = b = 0$
   (c) (i) $|a| = |b|$ and when $a = b = 0$ (ii) $a = b = 0$

Chapter 3

Linear Algebra

Areas of Interest

Equations (1)

Equations (2)

Equations (3)

Worded Problems

Equations and Inequations

Equations (1)

1. Solve the following equations for the pronumeral indicated.

(a) $5a - 4\frac{1}{2} = 3a + 7\frac{3}{4}$

(b) $5(9 - x) = 6(2x - 1)$

(c) $5 - 6x = 4(x + 5) - 3(2x - 11)$

(d) $\frac{2}{3 - 2x} = 1\frac{1}{3}$

(e) $\frac{x + 2}{3} = \frac{x + 4}{4}$

(f) $\frac{4x - 3}{12} = \frac{2x + 5}{4}$

(g) $\frac{5x - 1}{3} + \frac{2 - 3x}{4} = 2$

(h) $\frac{3x + 4}{5} + \frac{2x - 1}{3} = 3$

2. Solve the following equations.

(a) $\frac{-2}{5x} = \frac{1}{x} - 7$

(b) $\frac{x + 16}{x + 6} = \frac{x + 6}{x}$

(c) $\frac{4 + 6x}{1 - 4x} = -8$

(d) $\frac{5}{(x - 1)} - \frac{8}{(1 - x)} = 13$

Equations (1) page 2

3. For the following write an equation and find the value of x.
   (a) When five less than x is multiplied by five-eighths and has another x added, the result is 12.

   (b) When x less 7 is multiplied by negative 2 and then 6 is subtracted, the result is two-fifths of x.

   (c) When eight is added to x, the result multiplied by 3 and then 5 subtracted, the result is thirteen less than two-fifths of x.

   (d) When six is added to a twice x and the result is halved, the value is one less than a third of x.

Equations (2)

1. Find an expression for x in terms of a, b and c.

(a) $ax + c = bx$

(b) $a(bx - c) = b$

(c) $\frac{ax}{b} - c = ab$

(d) $bx - c = a(x + b)$

(e) $\frac{ax - b}{a + c} = bx$

(f) $\frac{ax}{b} + \frac{b}{c} = \frac{bx}{c} - a$

2. Solve for x.

(a) $\frac{1}{4} + \frac{1}{3} + \frac{1}{x} = 1$

(b) $\frac{1}{5} + \frac{1}{4} + \frac{1}{x} = 1$

(c) $-\frac{1}{4} + \frac{1}{3} - \frac{1}{2} + \frac{1}{x} = 1$

(d) $\frac{1}{4} - \frac{1}{3} + \frac{1}{2} - \frac{1}{x} = 1$

Equations (2) page 2

3. (a) (i) If $a = 36 - bc$, where a and b are variables and c is a constant, when $a = 12,\ b = 12$, find the value of b when $a = 18$.

(ii) Using the equation from (i), find the values of a and b, so that $a + b = 0$

(iii) Find the values of b so that $a \in [42, 54]$.

(b) If $ab = 36$, $\frac{b}{c} = \frac{1}{3}$, and $\frac{c}{a} = 27$, find the values of a, b and c that satisfy the equations.

Equations (3)

1. Find the values of a and b shown on the diagrams and hence, find the dimensions of the parallelograms.

(a)

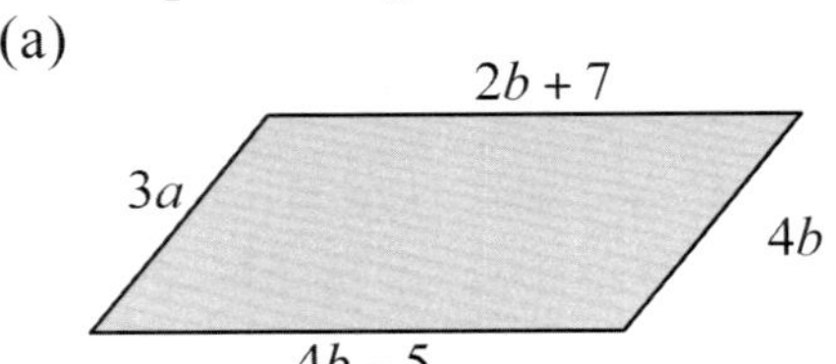

(b)

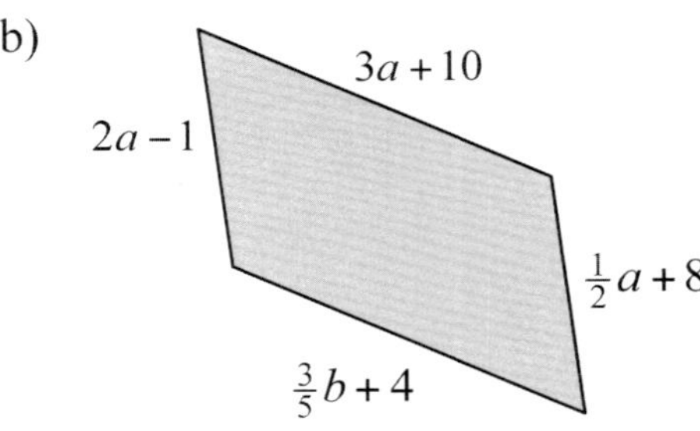

2. If $\frac{1}{2} > \frac{1}{3} > \frac{1}{a}$, and the difference between the first two fractions is the same as the difference between the last two fractions, find the value of a.

3. If $a = \dfrac{2b-8}{b+8}$, find the value of b, when $a = -4$.

Equations (3) page 2

4. Find an expression for x in terms of a and b.

(a) $\dfrac{a}{x+b} = \dfrac{b}{x+a}$

(b) $\dfrac{a}{x-a} = \dfrac{b}{x-b}$

(c) $\dfrac{-a}{x+b} = \dfrac{-b}{x+a}$

(d) $\dfrac{-a}{-x+b} = \dfrac{b}{x-a}$

5. If the perimeter of the shape is 56cm, find the value of a.
All angles are right angles.

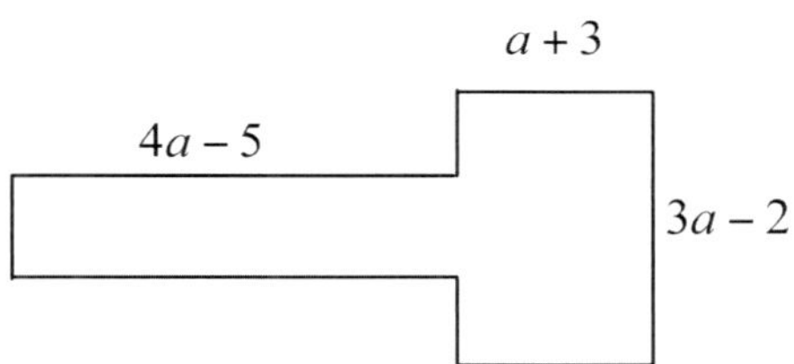

Worded Problems

Solve the following using algebra.

1. A water tank is $\frac{1}{a}$ full of water. When 12 litres are added, the tank is $\frac{1}{b}$ full. Find the capacity of the tank in terms of a and b.

2. A bluestone brick and a 2kg concrete block have the same weight as three-quarters of a bluestone brick and a four-and-a-half kilogram concrete block. Find the weight of $3\frac{2}{5}$ bluestone bricks and five, two-and-a-quarter kilogram concrete blocks.

3. The largest angle in a triangle is $27°$ greater than the smallest angle. If the smallest angle is $15°$ less than the third angle, find the difference between the largest angle and the third angle.

4. A blast furnace is operating at three-quarters its full temperature $(°C)$ when a power failure causes it to shut down. Two hours later the temperature has dropped $280°$ to 50% of its full operating temperature. Find the full operating temperature of the furnace and hence, find the temperature of the furnace one hour after the power failure if it cooled constantly.

5. Abe, Blair and Candy are three friends. Abe is ten years older than Blair and sixteen years older than Candy. The ratio of Abe's age to Blair's is the same as the ratio of Blair's age to Candy's age.
   (a) Find the ages of the friends and hence, find the ratio of their ages in 3 years time.

   (b) In how many years will Abe be twice as old as Candy?

Equations and Inequations

1. A rainwater tank initially full holds 18 000 litres of water. At midday, it begins to leak at the rate of 2 litres per minute.
   (a) Use V for volume of water remaining in the tank after t minutes and express the information as an equation.
   $V =$ ____________________

   (b) Find the volume of water remaining in the tank through the week at:
   (i) 12:43p.m. (ii) 2:15a.m.

   (c) The tank continues to leak constantly at this rate. Find the time it will take to empty in:
   (i) minutes (ii) hours (iii) days

   (d) If it started to leak at noon on Monday, find the day and the time that the volume of water remaining in the tank is:
   (i) 12 500 litres (ii) 1 250 litres

2. A washing machine repairman charges a fixed cost of \$90 plus \$30 for each 15 minute time period whilst fixing washing machines.
   (a) Using C for the charge made after t blocks of 15-minute time periods express the information as an equation.

   $C =$ ____________________

   (b) Find the charge made when repairs take the following amount of time:
   (i) 18 minutes (ii) 32 minutes (iii) 48 minutes (iv) 53 minutes

   (c) Express the time as an inequality for jobs that cost the following amounts:
   (i) \$120 (ii) \$240 (iii) \$330 (iv) \$90

Equations and Inequations page 2

3. A large apartment store is having a sale. Let P represent the original price for each of article and express the following as an equation and solve it to find the article's original price.

   (a) When a computer keyboard is discounted by 12% and is then further discounted by \$15, the cost is \$55.40.

   (b) A hair cutting kit has \$18 taken off the price and is then further discounted by 15% to a cost of \$37.40.

   (c) When a toaster is reduced by 20% and then is increased by \$12, the cost is \$71.80

   (d) A makeup kit's price is discounted by 10%, then reduced further by \$18 and then increased by 5% to a final cost of \$68.40

Solutions

Equations (1)

1. (a) $a = 6\frac{1}{8}$ (b) $x = 3$ (c) $x = -12$ (d) $x = \frac{3}{4}$ (e) $x = 4$
   (f) $x = -9$
   (g) $x = 2$ (h) $x = 2$
2. (a) $x = \frac{1}{5}$ (b) $x = 9$ (c) $x = \frac{1}{2}$
   (d) $x = 2$
3. (a) $x = 9\frac{4}{13}$ (b) $x = 3\frac{1}{3}$ (c) $x = -12\frac{4}{13}$ (d) $x = -6$

Equations (2)

1. (a) $x = \dfrac{c}{(b-a)}$ (b) $x = \dfrac{ac+b}{ab}$
   (c) $x = \dfrac{b(ab+c)}{a}$ (d) $x = \dfrac{ab+c}{(b-a)}$
   (e) $x = \dfrac{b}{(a-ab-bc)}$ (f) $x = \dfrac{b(b+ac)}{(b^2-ac)}$
2. (a) $x = 2\frac{2}{5}$ (b) $x = 1\frac{9}{20}$
   (c) $x = \frac{12}{17}$ (d) $x = -1\frac{5}{12}$
3. (a) (i) $b = 9$ (ii) $a = -36,\ b = 36$
   (iii) $b \in [-3, -9]$
   (b) $\{a, b, c\} = \{2, 18, 54\}, \{-2, -18, -54\}$

Equations (3)

1. (a) a = 8, b = 6, 19 units
   (b) a = 6, b = 40, 28 units
2. $a = 6$
3. $b = -4$
4. (a) $x = -(a+b)$
   (b)
   $x = 0$ and a and b can take any value
   or $a = b$ and x can take any value
   (c) $x = -(a+b)$
   (d) $= (a+b)$
5. $a = 4$

Worded Problems

1. $\dfrac{12ab}{(a-b)}$
2. $45\frac{1}{4}$kg
3. 12°
4. 700°
5. (a) 14:9:6 (b) 7 years

Equations and Inequations

1. (a) $V = 1800 - 2t$
   (b) (i) 12:43pm the same day: 17 914L, 12:43pm the next day: 15 034L
   12:43pm on the second day: 12 154L, 12:43pm on the third day: 9 274L
   12:43pm on the fourth day: 6 394L, 12:43pm on the fifth day: 3 514L
   12:43pm on the sixth day: 634L
   (ii) 2:15am the same day: 16 290L,
   2:15am the next day: 13 410L
   2:15am on the second day: 10 530L,
   2:15am on the third day: 7 650L
   2:15am on the fourth day: 4 770L,
   2:15am on the fifth day: 1 890L
   (c) (i) 9 000 minutes
   (ii) $\frac{9\,000}{60} = 150$ hours
   (iii) six-and-a-quarter days.
   (d)
   (i) Wednesday morning at 9:50am
   (ii) Sunday morning at 7:35am
2. (a) $C = 90 + 30t$
   (b) (i) \$150 (ii) \$180
   (iii) \$210 (iv) \$210
   (c) (i) $\{t : 0 \text{ minutes} \leq t \leq 15 \text{ minutes}\}$
   (ii) $\{t : 75 \text{ minutes} < t \leq 90 \text{ minutes}\}$
   (iii) $\{t : 120 \text{ minutes} < t \leq 135 \text{ minutes}\}$
   (iv) $t = 0$.
3. (a) \$80 (b) \$62 (c) \$74.75
   (d) \$92.38

Chapter 4

Linear Graphing

Areas of Interest

Equations of Straight Lines
Midpoint between Two Points
Gradient of Lines
Perpendicular Lines
Distance between Two Points
Points on Lines

Equations of Straight Lines

1. For each line shown on the set of axes below, find:
   (i) the gradient (ii) the y-intercept (iii) the equation.

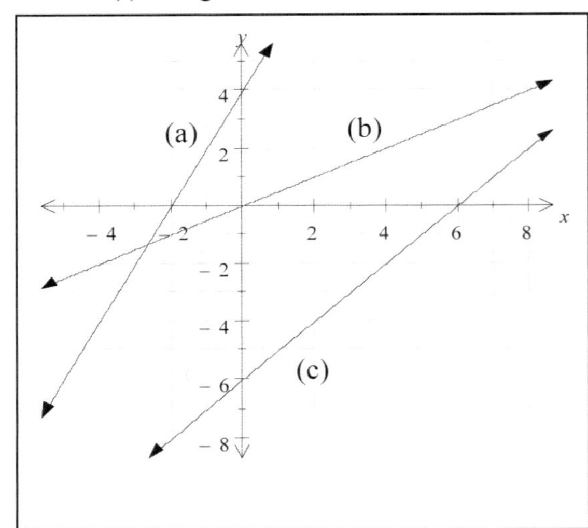

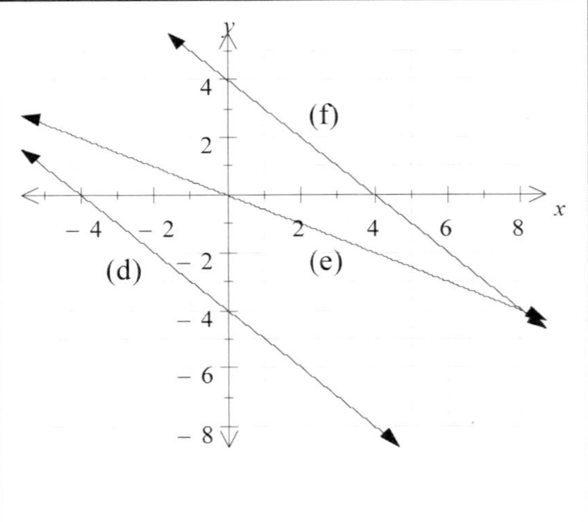

2. Find the equation of each of the lines shown on the axes.

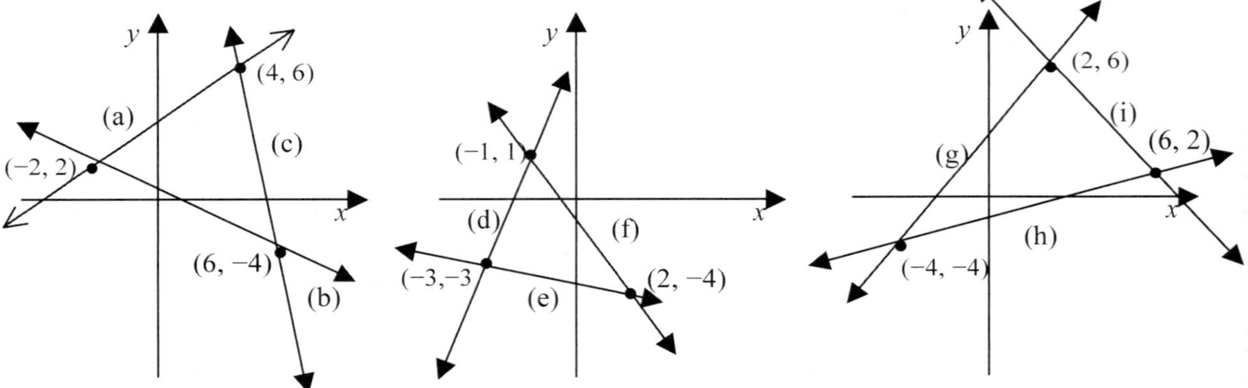

Equations of Straight Lines page 2

3. Find the equations of the lines using the clues below.
   (a) A horizontal line passing through the point (2, 4)

   (b) A vertical line passing through the point (−3, 5)

   (c) A line parallel to the line with equation $y - 3x + 4 = 0$, passing through the point (−9, 0)

   (d) A line parallel to the line with equation $x + y - 11 = 0$, passing through the point (4, −6)

4. The points (1, 2) and (3, 4) lie on the line $\frac{x}{a} + \frac{y}{b} = 2$. Find the values of a and b, and hence express the equation of the line in the form $y = mx + c$.

Midpoint between Two Points

1. (a) Plot the following points onto the set of axes and find the midpoint between each pair.

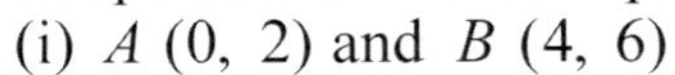
(i) A (0, 2) and B (4, 6)

(ii) C (−4, 0) and D (2, − 4)

(iii) E (−6, − 4) and F (−4, 6)

(iv) G (−3, 5) and H (3, − 3)

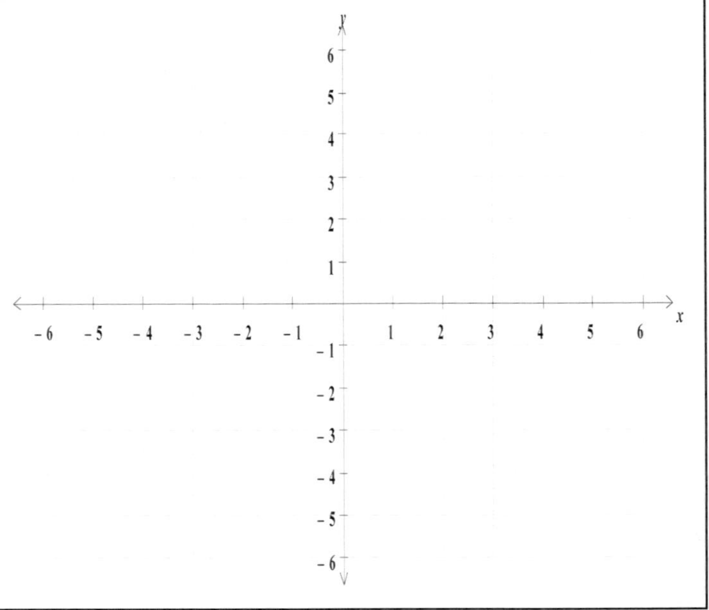

(b) Find a rule that can be used to find the midpoint between the general points (x_1, y_1) and (x_2, y_2).

(c) Use your rule to find the midpoint between the following.

(i) $(-a,\ b)$ and $(a,\ b)$

(ii) $(2a,\ -b)$ and $(4a,\ b)$

(iii) $(-a+1,\ b)$ and $(a+2,\ -b)$

(iv) $((a+b)^2,\ 2a^2)$ and $((a-b)^2,\ 2b^2)$

Midpoint between Two Points page 2

1. (d) Use the set of axes to show each line for the following. The point $C\ (-2, 2)$ is the midpoint of the line AB. Find the coordinates of B if the coordinates of A are:

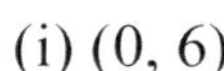
(i) (0, 6)

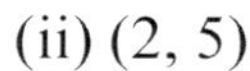
(ii) (2, 5)

(iii) (−6, 6)

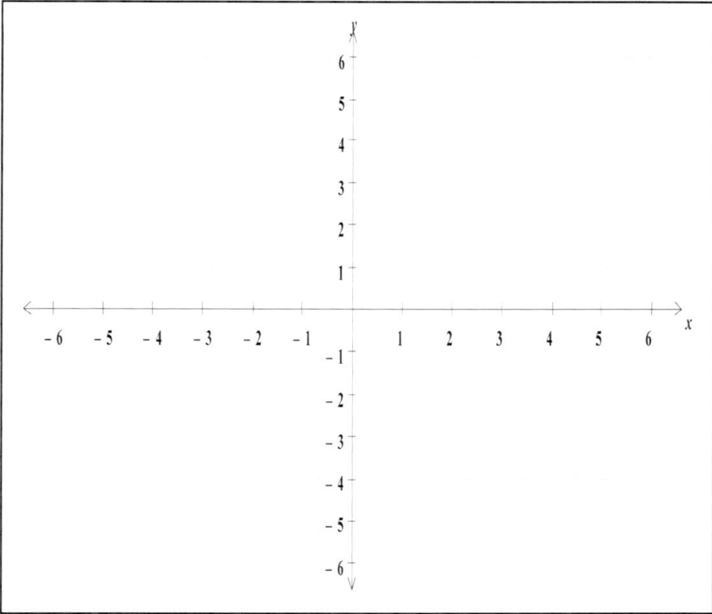

(e) Find a rule that can be used to find the coordinates of a second point (x_2, y_2), given that the coordinates of the first point are (x_1, y_1) and the coordinates of the midpoint between them are (x_m, y_m).

(f) Use your rule to find the coordinates of the second point, B if $A\ (x_1, y_1)$ and the midpoint of $AB\ (x_m, y_m)$ respectively are:

(i) $(-a,\ b)$ and $(a,\ b)$

(ii) $(2a,\ -b)$ and $(-4a,\ -3b)$

Gradient of Lines

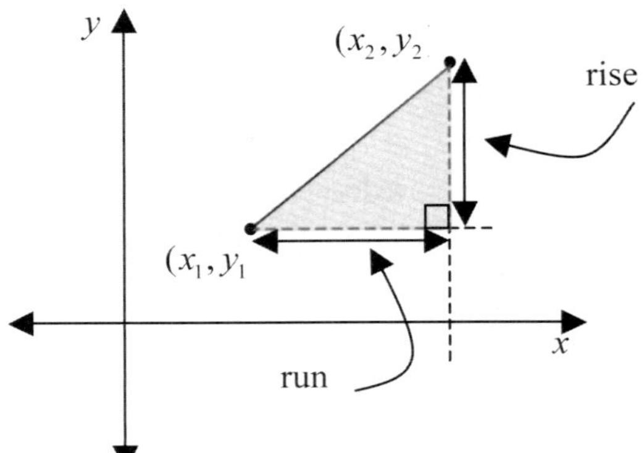

1. The gradient of a straight line (m) is a measure of its steepness. The gradient is given a numerical value, which is the ratio rise over run.

   $$\text{gradient } (m) = \frac{\text{rise}}{\text{run}}$$

   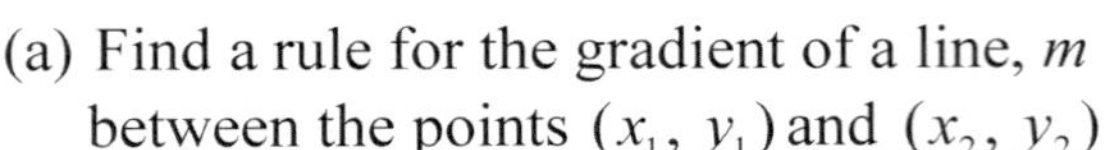

   (a) Find a rule for the gradient of a line, m between the points (x_1, y_1) and (x_2, y_2).

   (b) Use the rule to find the gradient between the following points.

   (i) (2, 4) and (8, 6)

   (ii) (−2, 2) and (2, 8)

   (iii) (−1, 5) and (−3, 9)

   (iv) (−2, −4) and (−4, 8)

   (v) (a, b) and (b, a)

   (vi) $(a, -b)$ and $(-a, b)$

2. By finding the gradients of the line segments, show that the following points lie on the same straight line and find the equation for each line.

   (a) $A\ (-2, 2)$, $B\ (2, 14)$ and $C\ (-5, -7)$

   (b) $A\ (-a, 4b)$, $B\ (a, 2b)$ and $C\ (2a, b)$

Gradient of Lines page 2

3. Show that the line segments AB and CD are parallel, where $A\ (-2,\ -3)$, $B\ (4,\ -2)$, $C\ (-1,\ -8)$ and $D\ (-7,\ -11)$.

4. The points A and B are joined. Find the value of a, so that the gradient of AB is:

(i) $\frac{1}{2}$ (ii) $-2\frac{1}{2}$

(a) $A\ (2,\ 4)$ and $B\ (a,\ 8)$ (b) $A\ (-4,\ a)$ and $B\ (6,\ -2)$

5. For $A\ (a,\ 4)$, $B\ (-2,\ 6)$, $C\ (-2,\ -4)$ and $D\ (-6,\ b)$, find the whole number values of a and b for $a \in [-10, 6]$ so that AB is parallel to CD.

Perpendicular Lines

1. Four lines are shown on the set of axes.
   (a) Find the equation of each line.

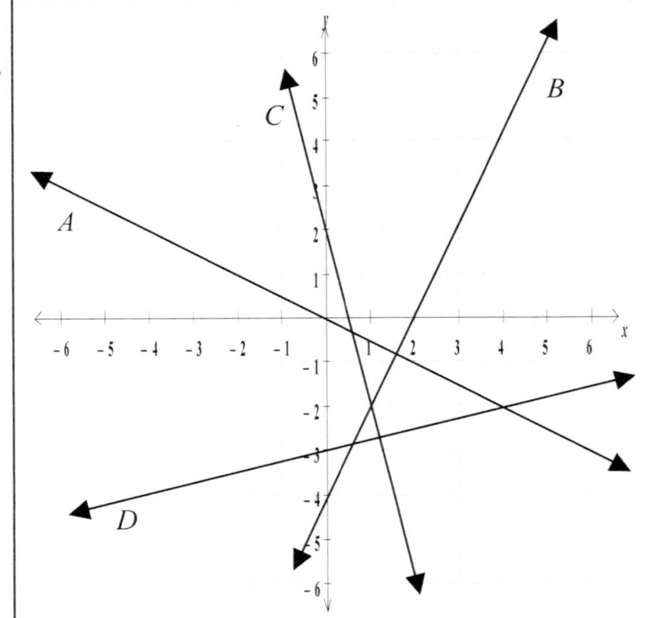

The lines *A*&*B* and *C*&*D* cross each other at right angles.

(b) What is special about the gradients of the lines:

(i) *A* and *B*?

(ii) *C* and *D*?

(c) Find a rule that connects the gradients of two lines, m_1 and m_2 that are perpendicular to each other.

(d) Use your rule to find the gradient of a line that is perpendicular to lines with equations:

(i) $y = -5x + 4$

(ii) $y = \frac{3}{4}x - 11$

(iii) $2y - x = 1$

(iv) $x - y + 1 = 0$

Perpendicular Lines page 2

2. The points $A\ (-5,\ -2)$, $B\ (-2,\ 4)$, C and D are vertices of a square.

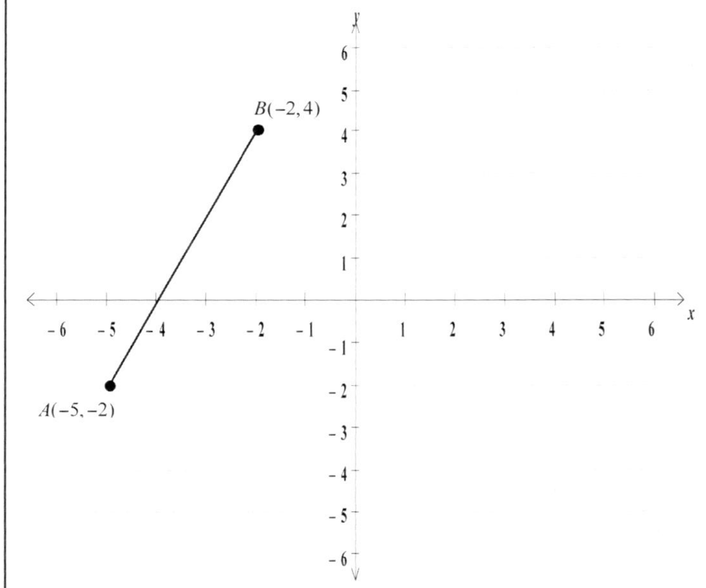

(a) Find the coordinates of the point C, which is in the first quadrant and hence, find the equation of BC.

(b) Find the coordinates of the point D and hence, find the equation of the lines CD and AD.

(c) Use the line AB to find the coordinates of C and D to make another square, hence find the equation of its side lengths.

3. Show that the triangle ABC is right-angled for $A\ (2,\ -3)$, $B\ (5,\ 2)$ and $C\ (-3,\ 0)$.

Distance between Two Points

1. The line segments AB and CD are shown on the set of axes.
   (a) For AB a right-angled triangle is shown.
   (i) Use the scale to find the height and base length of the triangle.

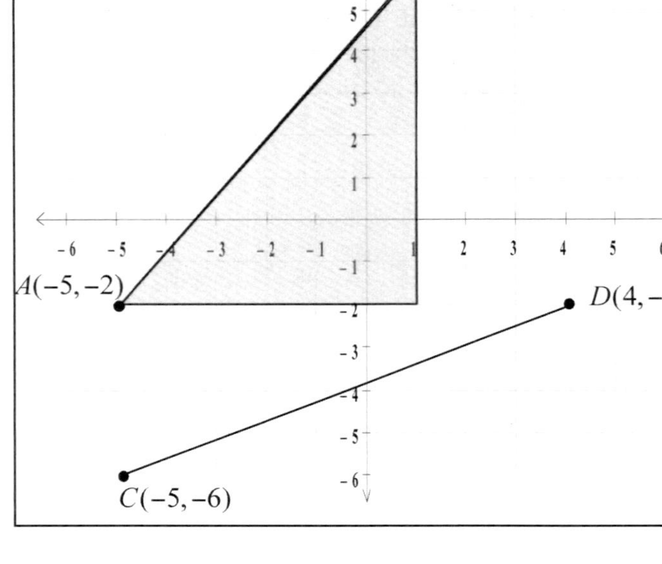

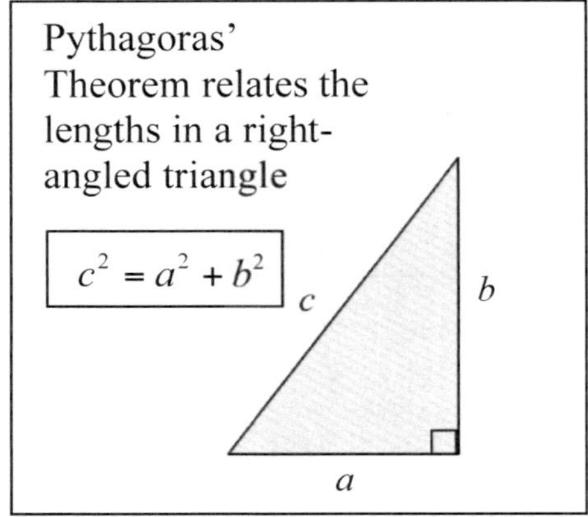

(ii) Use Pythagoras' Theorem to find the length of AB.

(iii) Find the length of CD expressed in the form of a square root (surd).

(b) The points A and B are shown on the graph.
   (i) Place the values x_1, y_1, x_2 and y_2 on the axes.

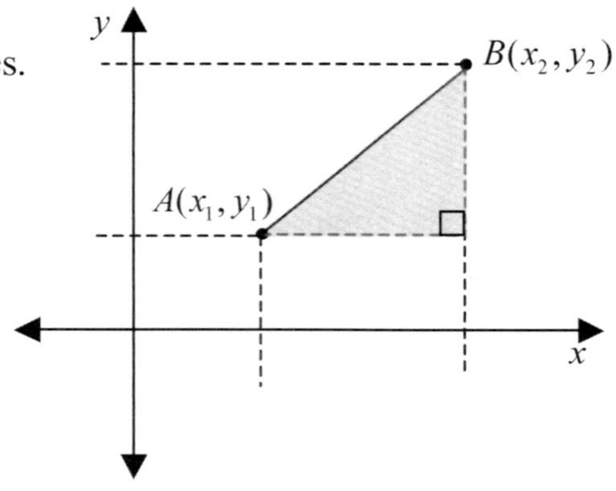

Distance between Two Points page 2

1. (b) (ii) Find a rule that expresses the distance AB in terms of x_1, y_1, x_2 and y_2 using Pythagoras's theorem.

   (iii) Use your rule to find the distance between the points $A\ (2,\ 5)$ and $B\ (5,\ 9)$.

2. Show that the point $A\ (10,\ 7)$ is the same distance from the points $B\ (3,\ 6)$ and $C\ (5,\ 2)$.

3. (a) Find the coordinates of C, which is the midpoint of the interval AB, for $A\ (a,\ 2b)$ and $B\ (3a,\ 4b)$.

   (b)Find the distances of the intervals AC and BC, hence prove that C is the midpoint of AB.

Points on Lines

1. Show that the following points lie on the lines with equations:

   (a) $y = \frac{1}{4}(x - 45)$ (i) $(1,\ -11)$ (ii) $(-1,\ -11\frac{1}{2})$ (iii) $(-5,\ -12\frac{1}{2})$

   (b) $\dfrac{x + y}{4} = -1$ (i) $(-5\frac{1}{4},\ 1\frac{1}{4})$ (ii) $(-3\frac{1}{2},\ -\frac{1}{2})$ (iii) $(1\frac{1}{4},\ -5\frac{1}{4})$

   (c) $2a - y = 3(x + b)$ (i) $(-b,\ 2a)$ (ii) $(-a,\ 5a - 3b)$ (iii) $(a - b,\ -a)$

2. (a) Find the equation of the line passing through the points $A\ (a,\ b)$ and $B\ (a + 1,\ b - 1)$.

   (b) Show that the point $(a - 1,\ b + 1)$ lies on AB.

   (c) Find the gradient of a line, which is perpendicular to AB, passing through (i) A (ii) B and hence, state the difference in position of the lines.

3. The line AB has the equation $y = 2x + 8$.

   (a) If the points A and B have coordinates $(-2,\ 4)$ and $(4,\ a)$, find the value of a.

Points on Lines page 2

3. (b) The point C lies on the line which is perpendicular to AB through B with coordinates $(8,\ b)$. Find the value of b.

(c) Find the area of the triangle ABC.

4. (a) Find the point of intersection $A\ (x,\ y)$ between the lines with equations $x-3y=8$ and $2x-y-6=0$.

(b) If the point A lies on the line $kx+5y-2=0$, find the value of k, hence find the axis intercepts of the line.

Solutions

Equations of Straight Lines

1. (a) (i) $m = \frac{4}{2} = 2$ (ii) $c = 4$
   (iii) $y = 2x + 4$
   (b) (i) $m = \frac{1}{2}$ (ii) $c = 0$
   (iii) $y = \frac{1}{2}x$
   (c) (i) $m = \frac{6}{6} = 1$ (ii) $c = -6$
   (iii) $y = x - 6$
   (d) (i) $m = -\frac{4}{4} = -1$ (ii) $c = -4$
   (iii) $y = -x - 4$
   (e) (i) $m = -\frac{1}{2}$ (ii) $c = 0$
   (iii) $y = -\frac{1}{2}x$
   (f) (i) $m = -\frac{4}{4} = -1$ (ii) $c = 4$
   (iii) $y = -x + 4$
2. (a) $y = \frac{2}{3}x + 3\frac{1}{3}$ (b) $y = -\frac{3}{4}x + \frac{1}{2}$
   (c) $y = -5x + 26$ (d) $y = -x + 3$
   (e) $y = -\frac{1}{5}x - 3\frac{3}{5}$
   (f) $y = -1\frac{2}{3}x + \frac{2}{3}$
   (g) $y = 1\frac{2}{3}x + 2\frac{2}{3}$
   (h) $y = \frac{3}{5}x - 1\frac{3}{5}$ (i) $y = -x + 8$
3. (a) $y = 4$ (b) $x = -3$
   (c) $y = 3x + 27$ (d) $y = -x - 2$
4. $a = -\frac{1}{2}$, $b = \frac{1}{2}$ and $y = x + 1$

Midpoint between Two Points

1. (a) i) (2, 4) (ii) (−1, −2)
   (iii) (−5, 1) (iv) (0, 1)
   (b) $(x_m, y_m) = \left(\frac{x_1 + x_2}{2}, \frac{y_1 + y_2}{2}\right)$
   (c) (i) $(0, b)$ (ii) $(3a, 0)$
   (iii) $(1\frac{1}{2}, 0)$ (iv) $(a^2 + b^2, a^2 + b^2)$
   (d) (i) (−4, −2) (ii) (−6, −1)
   (iii) (2, −2)
   (e) $(2x_m - x_1, 2y_m - y_1)$
   (f) (i) $(3a, b)$ (ii) $(-10a, -5b)$

Gradient of Lines

1. (a) Gradient $(m) = \dfrac{(y_2 - y_1)}{(x_2 - x_1)}$
   (b) (i) $\frac{1}{3}$ (ii) $1\frac{1}{2}$ (iii) −2
   (iv) −6 (v) −1 (vi) $-\frac{b}{a}$
2. (a) $m(AB) = m(BC) = 3$
   (b) $m(AB) = m(BC) = -\frac{b}{a}$
3. $m(AB) = m(CD) = \frac{1}{2}$
4. (a) (i) $a = 10$ (ii) $a = \frac{2}{5}$
   (b) (i) $a = 23$ (ii) $a = -7$
5.

a	b
−10	−5
−6	−6
−4	−8
−3	−12
−1	4
0	0
2	−2
6	−3

Perpendicular Lines

1. (a)
   Line A:
   $m = -\frac{1}{2}$, $c = 0$ Equation: $y = -\frac{1}{2}x$
   Line B:
   $m = 2$, $c = -4$ Equation: $y = 2x - 4$
   Line C:
   $m = -4$, $c = 2$ Equation: $y = -4x + 2$

   Line D:
   $m = \frac{1}{4}$, $c = -3$ Equation: $y = \frac{1}{4}x - 3$
   (b) The gradient of one line is the negative reciprocal of the gradient of the other line, or the product of the gradients is negative one.

1. (c) $m_1 \times m_2 = -1$
   (d) (i) -5 (ii) $\frac{3}{4}$
   (iii) $\frac{1}{2}$ (iv) 1
2. (a) $C\ (4,\ 1)$, $y = -\frac{1}{2}x + 3$
   (b) $D\ (1,\ -5)$
   CD: $y = 2x - 7$
   AD: $y = -\frac{1}{2}x - 4\frac{1}{2}$
   (c) $C\ (-8,\ 7)$, $D\ (-11,\ 1)$.
   BC: $y = -\frac{1}{2}x + 3$
   AD: $y = -\frac{1}{2}x - 4\frac{1}{2}$
   CD: $y = 2x + 23$

Distance between Two Points

1. (a) (i) Base: 6 units, Height: 8 units
   (ii) 10 units (iii) $\sqrt{97}$ units

   (b) (ii)

Length $AB = \sqrt{(x_2 - x_1)^2 + (y_2 - y_1)^2}$

   (iii) 5 units

3. (a) $(2a,\ 3b)$
   (b)

Length AC = Length $BC = \sqrt{a^2 + b^2}$

Points on Lines

2. (a) $y = -x + (a + b)$ (c) 1
3. (a) $a = 16$ (b) $b = 14$
   (c) 30 square units
4. (a) $A = (2, -2)$ (b) $k = 6$,

The x-intercept: $x = \frac{1}{3}$, y-intercept: $y = \frac{2}{5}$

Chapter 5

Geometry

Areas of Interest

Angles (1)

Angles (2)

Circle Geometry

Shapes and Geometry

Angles (3)

Angles (1)

1. Find the angle that the hour hand turns every:
   (a) two hours
   (b) x hours
   (c) $\frac{x}{3}$ of an hour
   (d) $\frac{n}{m}$ of an hour, for $m = 6n$

2. Find the value of x for the following.

(a)

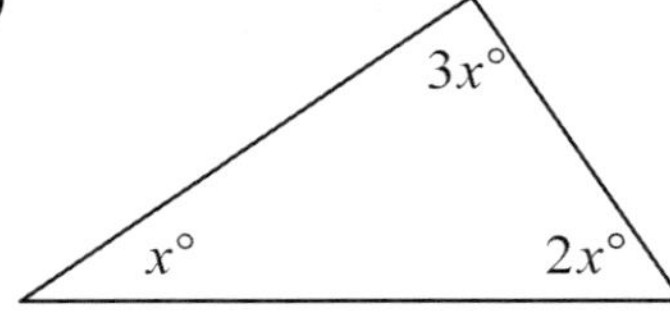

(b)

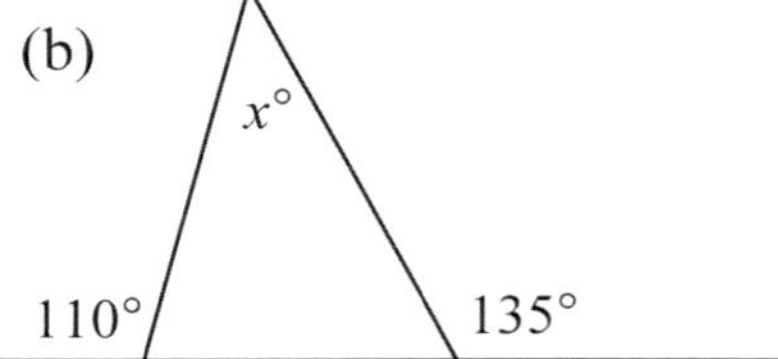

(c)

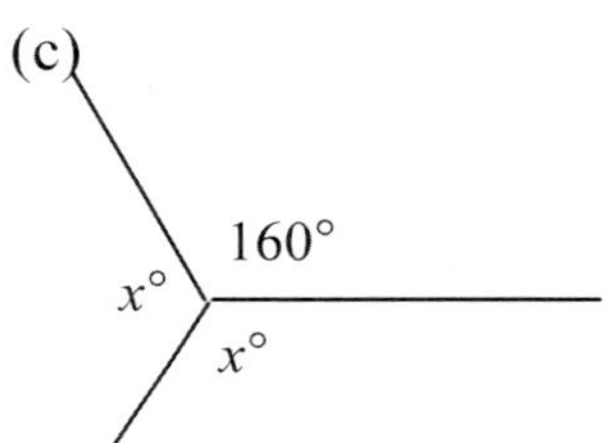

(d)

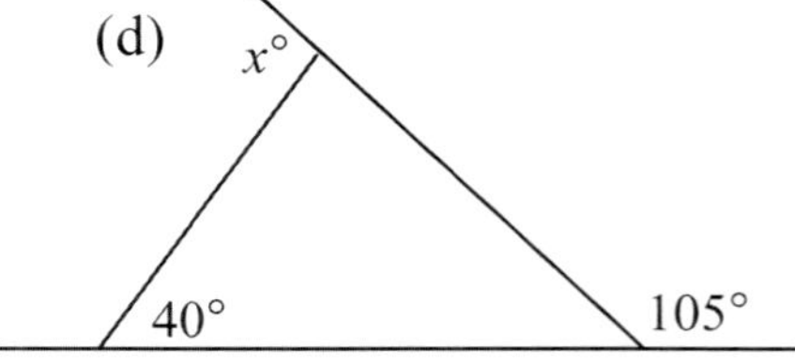

(e)

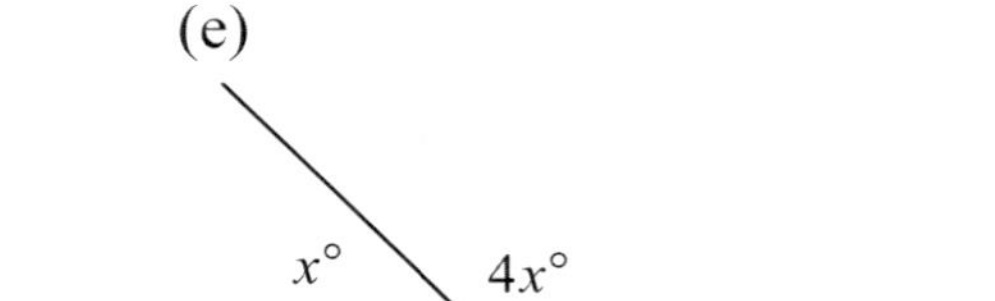

(f)

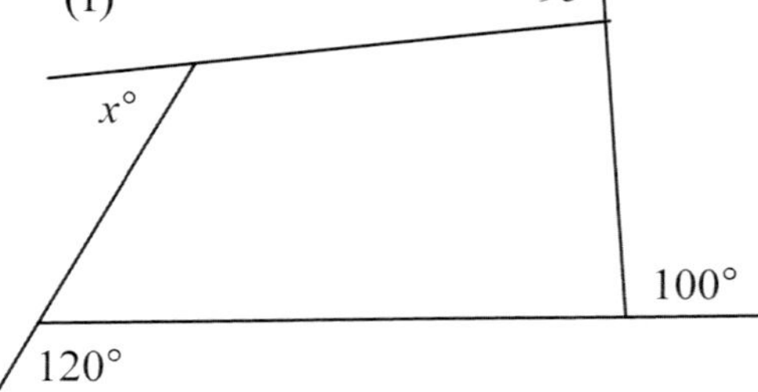

Angles (1) page 2

2. (g)

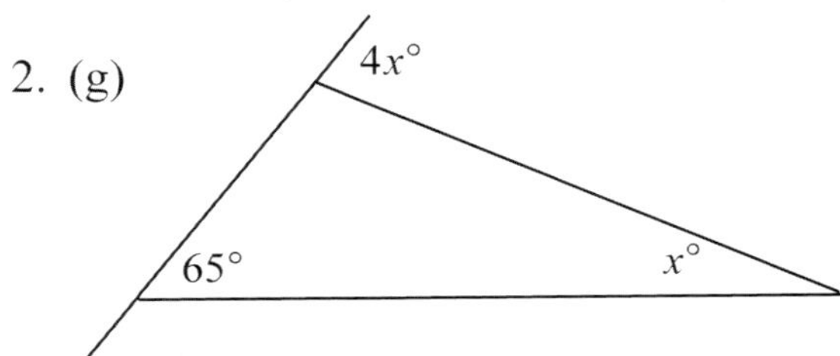

(h)

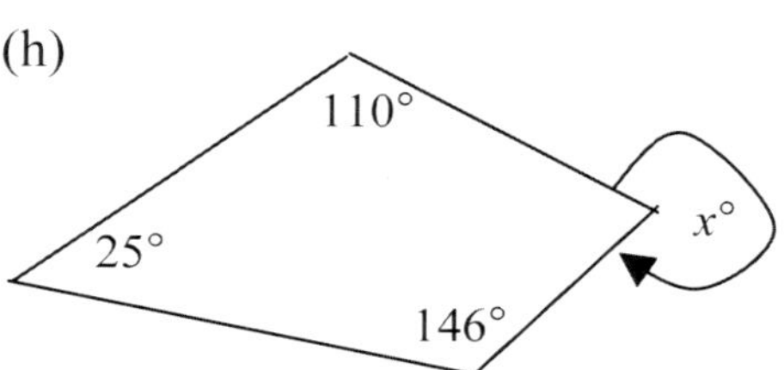

3. If the angles in a triangle are in the following ratios, find the difference between the largest and smallest angle.

(a) 2 : 3 : 4 (b) 1 : 4 : 5 (c) 4 : 5 : 6 (d) 1 : 7 : 7

4. Find the smaller angle between the hands of a clock at:

(a) 3 : 30 (b) 12 : 30 (c) 1:15 (d) 4 : 20

Angles (2)

1. In the diagram $\angle ABF = \angle DBF$ and $\angle EDF = \angle BDF$.
   Find $\angle ACE$ if:
   (a) $\angle ABF = 40°$ and $\angle BFD = 68°$

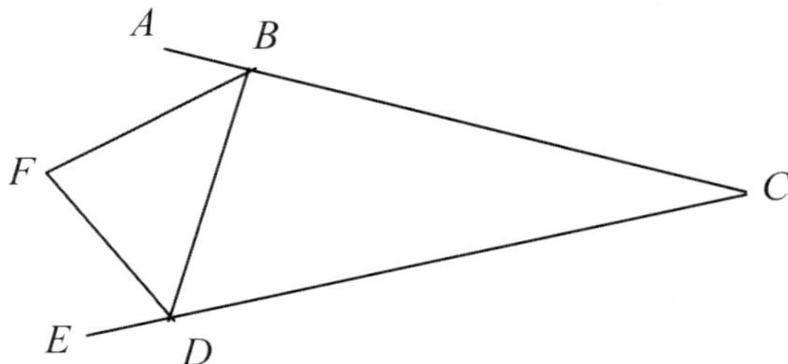

   (b) $\angle BFD = 80°$ and $\angle ABF = 30°$

   (c) $\angle BFD = 60°$ and $\angle ABF = 2\angle EDF$

2. In the diagram $AB = OC = r$ and AD passes through the centre O of the circle. Find $\angle OAC$ if $\angle COD = 81°$.

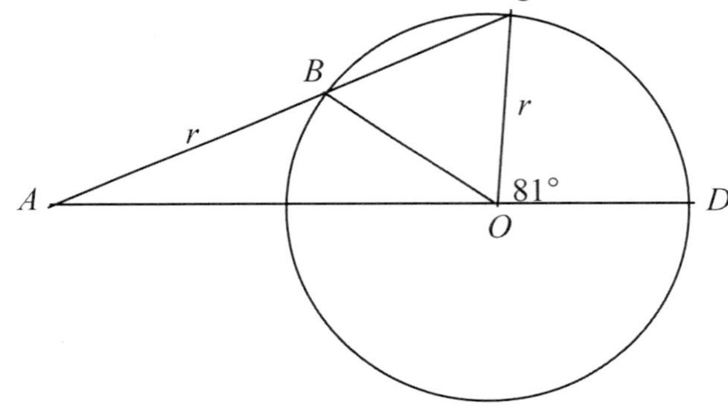

Angles (2) page 2

3. In this diagram, there are four sets of parallel lines. Triangle ABC is an equilateral triangle where $dAD = dCD$.
   Find the angles: (i) a (ii) b (iii) c (iv) d (v) e (vi) f (vii) g (viii) h (ix) i

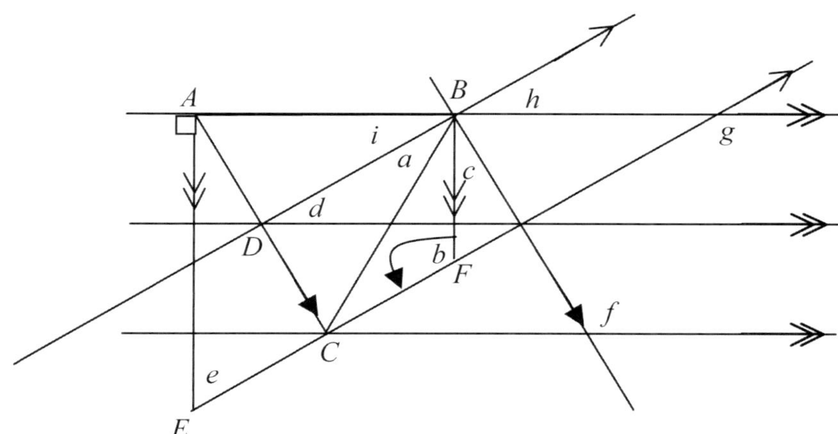

Circle Geometry

1. Find the value of the pronumerals. O is the centre of the circle.

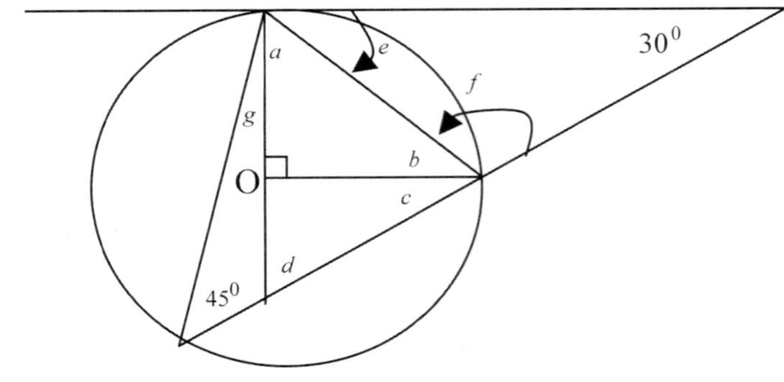

$a =$ ________ $b =$ ________

$c =$ ________ $d =$ ________

$e =$ ________ $f =$ ________

$g =$ ________

2. Find the value of the pronumerals. O is the centre of the circle

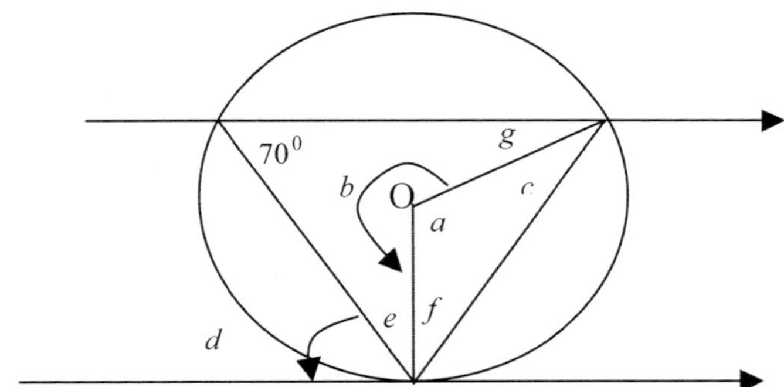

$a =$ _______ $b =$ ________

$c =$ _______ $d =$ ________

$e =$ _______ $f =$ ________

$g =$ _______

Circle Geometry page 2

3. Find the value of the pronumerals. O is the centre of the circle:

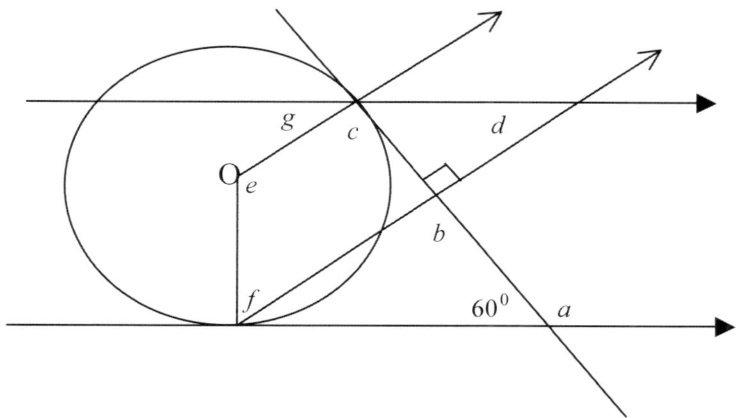

$a =$ _______ $b =$ _______

$c =$ _______ $d =$ _______

$e =$ _______ $f =$ _______

$g =$ _______

4. The diagram below features a circle with centre ●. Find the value of the pronumerals.

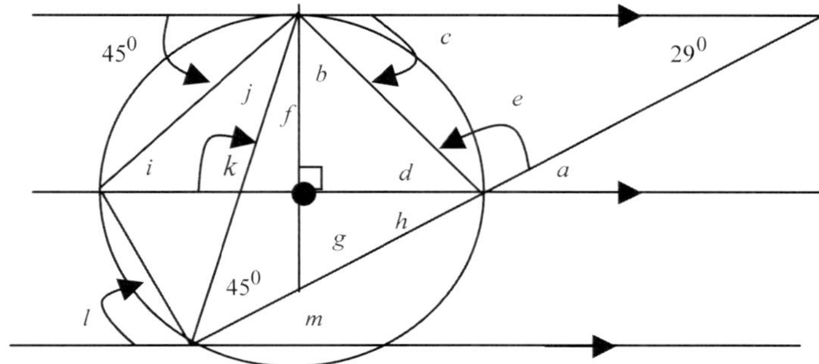

$a =$ _____ $b =$ _____ $c =$ _____

$d =$ _____ $e =$ _____ $f =$ _____

$g =$ _____ $h =$ _____ $i =$ _____

$j =$ _____ $l =$ _____ $m =$ _____

Shapes and Geometry

1. For the parallelogram $ABCD$, draw four squares using the edges of the parallelogram.

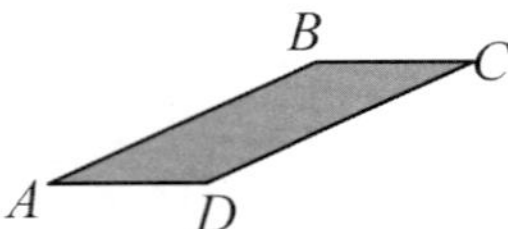

Locate the centres of the squares by drawing in the diagonals and join them.
(a) State the shape that is formed.

(b) Start with another shaped parallelogram and see if the same result happens.

2. (a) For the quadrilateral $ABCD$, join the midpoints of edges and name the shape that results.

(b) Draw another shaped quadrilateral and examine the nature of the shape.

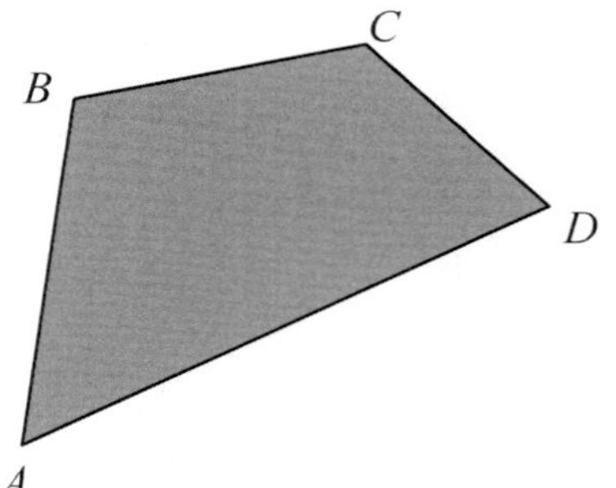

Shapes and Geometry page 2

3. For the triangle *ABC*, draw three equilateral triangles using the edges of the triangle.

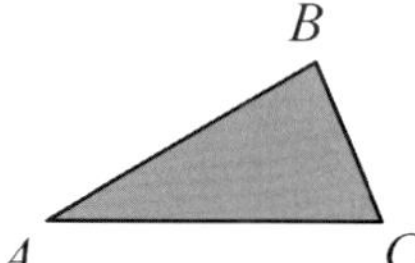

Locate the centres of the equilateral triangles by drawing in the medians (a median is a line drawn from each vertex to the centre of the opposite edge). Join the centres of the equilateral triangles.

(a) State the shape that is formed.

(b) Start with another shaped triangle and see if the same result happens.

4. Three equilateral triangles *ABC*, *CDE* and *AEF* are shown. Locate the centres of the equilateral triangles by drawing in the medians (a median is a line drawn from each vertex to the centre of the opposite edge). Join the centres of the equilateral triangles.

   (a) State the shape that is formed.

   (b) Move the point *E* along *AC* to another position and draw the new equilateral triangles. Join the centres of the triangles and examine the nature of the triangle.

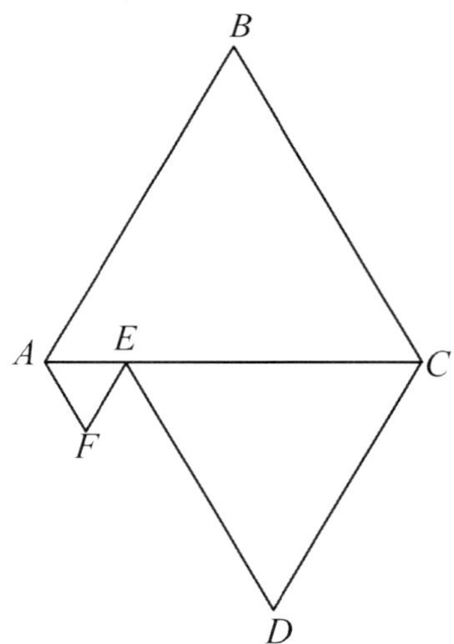

Angles (3)

1. (a) Find the sum of the given angles in the diagrams.

(i)

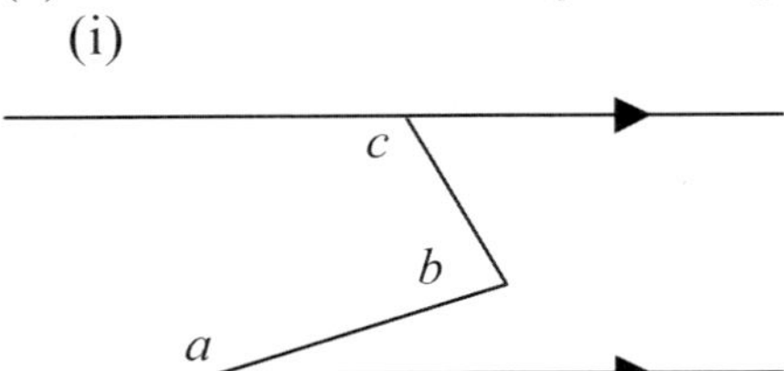

(ii)

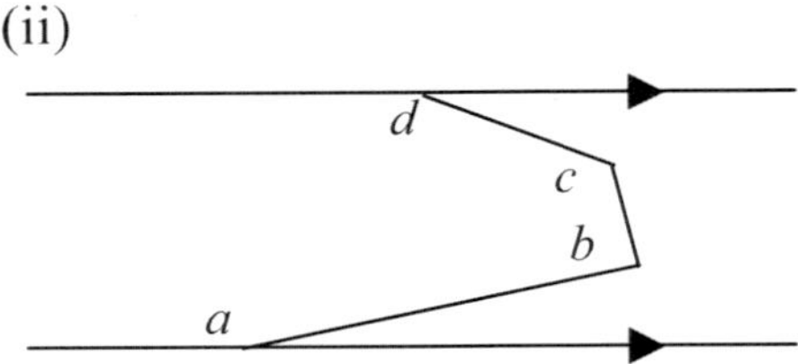

(b) Use the results above to find the angle sum when n angles are constructed between the parallel lines.

2. (a) In the diagram AC is the diameter of the circle, $\angle AEB = \alpha$. Express all the other angles in terms of α .

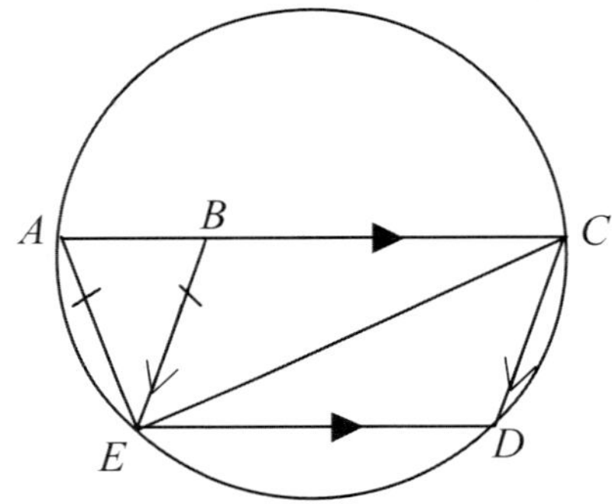

(b) In the diagram $\angle ABD = \angle ADC = 90°$ and $\angle CAD = \theta$. Express all the angles in terms of θ .

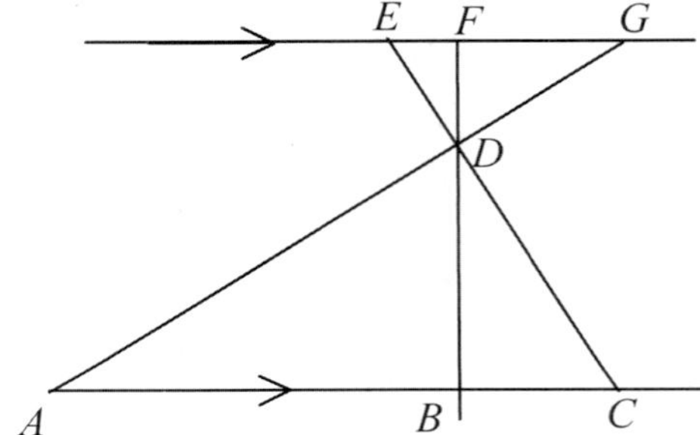

Angles (3) page 2

3. (a) (i) Find the size of all the angles in the diagram if $ABCDE$ is a regular pentagon, CF is a diameter of the circle and $AG = EG$.

(ii) Hence, state which triangles are similar or isosceles.

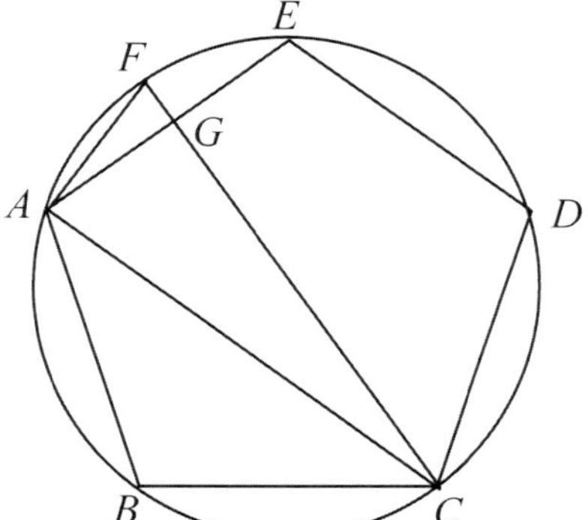

(b) (i) Find the size of all the angles in the diagram if $ABCDEF$ is a regular hexagon.

(ii) Hence, state which triangles are similar, congruent or isosceles.

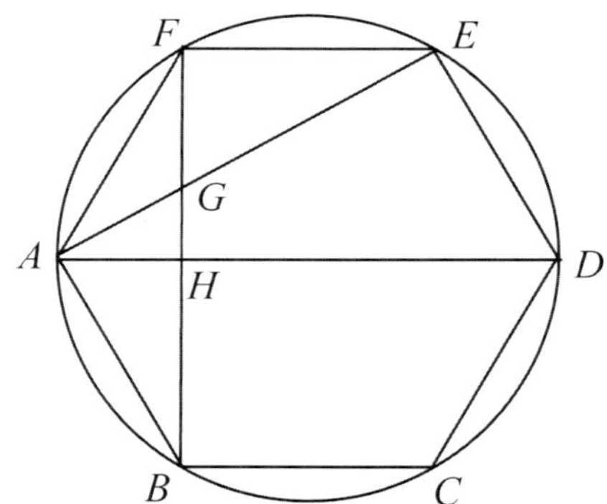

Solutions

Angles (1)

1. (a) 60° (b) $30x^\circ$ (c) $10x^\circ$ (d) 5°
2. (a) 30° (b) 65° (c) 100° (d) 115°
   (e) 36° (f) 45° (g) $21\frac{2}{3}^\circ$ (h) 281°
3. (a) 40° (b) 72° (c) 24° (d) 72°
4. (a) 75° (b) 165° (c) $82\frac{1}{2}^\circ$ (d) 10°

Angles (2)

1. (a) 36° (b) 20° (c) 60°
2. 27°
3. (i) 30° (ii) 120° (iii) 30° (iv) 30° (v) 60° (vi) 120° (vii) 150° (viii) 30° (ix) 30°

Circle Geometry

1. $a = b = 45^\circ$, $e = 45^\circ$, $d = 60^\circ$, $c = 30^\circ$, $g = 15^\circ$, $f = 105^\circ$
2. $a = 140^\circ$, $b = 220^\circ$, $c = f = 20^\circ$, $d = 70^\circ$, $e = g = 20^\circ$
3. $a = 120^\circ$, $b = 90^\circ$, $c = 90^\circ$, $e = 120^\circ$, $d = 30^\circ$, $f = 60^\circ$, $g = 30^\circ$
4. $a = 29^\circ$, $b = d = i = 45^\circ$, $c = 45^\circ$, $e = 106^\circ$, $f = 16^\circ$ $h = 29^\circ$, $g = 61^\circ$, $m = 29^\circ$, $l = 61^\circ$, $k = 106^\circ$, $j = 29^\circ$

Shapes and Geometry

1. (a) Square
2. (a) Parallelogram
3. (a) Equilateral triangle
4. (a) Equilateral triangle

Angles (3)

1. (a) (i) $2 \times 180^\circ$ (ii) $3 \times 180^\circ$
   (b) $(n-1) \times 180^\circ$
2. (a) $\angle AEC = 90^\circ$, $\angle BEC = 90^\circ - \alpha$, $\angle ECD = 90^\circ - \alpha$, $\angle BAE = \angle EBA = 90^\circ - \frac{1}{2}\alpha$, $\angle CBE = 90^\circ + \frac{1}{2}\alpha$, $\angle CDE = 90^\circ + \frac{1}{2}\alpha$, $\angle CED = \frac{1}{2}\alpha$, $\angle BCE = \frac{1}{2}\alpha$
   (b) $\angle ABD = \angle CBD = \angle EFD = \angle GFD = 90^\circ$, $\angle EDG = \angle EDA = \angle CDG = 90^\circ$, $\angle ADB = 90^\circ - \theta$, $\angle ACD = 90^\circ - \theta$, $\angle CDB = \theta$, $\angle EDF = \theta$, $\angle FDG = 90^\circ - \theta$, $\angle FGD = \theta$, $\angle CEG = 90^\circ - \theta$
3. (a) (i) $\angle ABC = \angle BCD = \angle CDE = \angle DEA = \angle EAB = 108^\circ$, $\angle BAC = \angle ACB = 36^\circ$, $\angle CAF = 90^\circ$, $\angle CAG = 72^\circ$, $\angle CGA = 90^\circ$, $\angle AGF = 90^\circ$, $\angle ACG = 18^\circ$, $\angle AFC = 72^\circ$, $\angle DCG = 54^\circ$, $\angle EGC = \angle AGF = 90^\circ$, $\angle GAF = 18^\circ$
   (ii) Isosceles: $\triangle_{ABC}$, Similar triangles: $\triangle_{ACG}$ and $\triangle_{AFG}$

3. (b) (i) $\angle ABC = \angle BCD = \angle CDE = \angle DEA = \angle EFA = \angle FAB = 120°$,
$\angle CDA = \angle EDA = \angle BAD = 60°$, $\angle CBH = \angle HFE = 90°$, $\angle AHB = 90°$,
$\angle AHB = \angle BHD = \angle AHF = \angle DHF = 90°$, $\angle ABH = 30°$, $\angle BAH = 60°$,
$\angle AFG = 30°$, $\angle FAG = \angle AFG = 30°$, $\angle AGF = 120°$, $\angle EGH = \angle AGF = 120°$,
$\angle EAD = 30°$, $\angle ADE = 60°$, $\angle AGH = 60°$, $\angle EGH = 60° = \angle AGH$, $\angle FEG = 30°$,
$\angle AEF = 90°$

(ii)

Congruent triangles are the group $\triangle_{ABH}, \triangle_{AFH}, \triangle_{EFG}$ and the group $\triangle_{ABF}, \triangle_{AEF}$

Similar triangles: $\triangle_{AFG}, \triangle_{AEF}, \triangle_{ABF}$ and $\triangle_{ABH}, \triangle_{AFH}, \triangle_{EFG}, \triangle_{AGH}$

Isosceles triangles: $\triangle_{AFG}, \triangle_{AEF}, \triangle_{ABF}$

Chapter 6

Indices

Areas of Interest

Index Rules and Algebra

Indices and Number

Indices and Algebra (1)

Indices and Algebra (2)

Index Rules and Algebra

1. (a) Express the following to nine decimal places.

   (i) $3^{\frac{1}{2}}$ ________________ $\sqrt{3}$ ________________

   (ii) $5^{\frac{1}{2}}$ ________________ $\sqrt{5}$ ________________

   (b) Use the results found above to make a statement between the expressions $a^{\frac{1}{2}}$ and $\sqrt{a}$.

   (c) Show this result to be true for all values by squaring both expressions.

2. (a) Evaluate the following using a calculator.

   (i) 2^0 ________ (ii) $(-23)^0$ ________

   (b) Simplify the following.

   (i) $\frac{a^2}{a^2} =$ ________ (ii) $\frac{a^2}{a^2} = a^{2-2} =$ __________

   (c) Use the above to explain why $a^0 = 1$.

3. (a) Express the following in decimal form and as fractions.

   (i) 2^{-1} _______ (ii) 4^{-1} ________

   (b) Simplify the following.

   (i) $a^3 \div a^4$ (ii) $\frac{a^3}{a^4}$

   (c) Use the above to explain why $a^{-1} = \frac{1}{a}$.

Index Rules and Algebra page 2

4. Fill in the gaps to show how the following are calculated.

(a)

$5^3 = 5 \times 5^2$

$= 5 \times 25$

$= 4 \times 25 + 1 \times 25$

$= (2 \times \_\_)^2 + (\_\_\_)^2$

$= (\_\_\_)^2 + (\_\_\_)^2$

$= \_\_\_\_\_$

(b) Use the result to show why $a^2 + b^2 \neq (a+b)^2$.

5. Complete the following number pattern.

$(1+2)^2 - 1^2 = \_\_\_\_\_\_\_\_\_\_ = \_\_\_\_ = (\_\_\_\_)$

$(1+2+3)^2 - (1+2)^2 = \_\_\_\_\_\_\_\_\_\_\_\_ = \_\_\_\_ = (\_\_\_\_)$

$(1+2+3+4)^2 - (1+2+3)^2 = \_\_\_\_\_\_\_\_\_\_\_\_ = \_\_\_\_ = (\_\_\_\_)$

$(1+2+3+4+5)^2 - (1+2+3+4)^2 = \_\_\_\_\_\_\_\_\_\_\_\_ = \_\_\_\_ = (\_\_\_\_)$

$(1+2+3+.....+n)^2 - (1+2+3+....(n-1))^2 = (\_\_\_\_)$

Indices and Number

1. Show how:

(a) $8^{\frac{2}{3}} = 4$

(b) $16^{-\frac{3}{4}} = \frac{1}{8}$

2. Simplify.

(a) $\left(9\times5^3\right)^{\frac{1}{2}}\times\left(4\times5\right)^{-1\frac{1}{2}}$

(b) $\frac{1}{\sqrt{2}}\times2^{\frac{1}{2}}\times\frac{2}{\sqrt{2}}$

3. Evaluate the following without using a calculator and show full working.

(a) $16^{\frac{1}{4}}$

(b) $625^{\frac{3}{4}}$

(c) $4^{\frac{3}{2}}$

(d) $8^{\frac{7}{3}}$

(e) $32^{\frac{2}{5}}$

(f) $121^{-\frac{1}{2}}$

(g) $27^{-\frac{5}{3}}$

(h) $125^{-\frac{2}{3}}$

4. Express the following with a prime number base.

(a) 27^2

(b) 8×2^a

(c) 27×3^b

(d) $5^a\div5$

Indices and Number page 2

4. (e) $2^a \times 4^a$ (f) $\frac{16}{2^a}$

(g) $\frac{2^{a+1}}{2^{1-a}}$ (h) $\frac{27^a}{9^b}$

5. Use indices to evaluate the following, show all working.

(a) $125^{\frac{1}{3}}$ (b) $(8 \times 27)^{\frac{2}{3}}$

(c) $2401^{\frac{1}{2}} \times 625^{-\frac{1}{2}}$ (d) $\left(8^{\frac{1}{3}} \times 16^{\frac{1}{4}}\right)^2$

6. Simplify.

(a) $\frac{75^{\frac{1}{2}} \times 125}{9^{\frac{1}{4}} \times 5^{-1}}$ (b) $\frac{24^{\frac{1}{2}} \times 48}{12^{\frac{1}{4}} \times 3^{-\frac{1}{3}}}$

Indices and Algebra (1)

1. Complete a summary of the laws of indices.

(a) $x^a \times x^b =$ (b) $\dfrac{x^a}{x^b} =$ (c) $\left(x^a\right)^b =$

(d) $\sqrt[a]{x} =$ (e) $a^0 =$ (f) $x^{-a} =$

2. Fully simplify the following expressions.

(a) $\left(n^{\frac{1}{2}}\right)^5$ (b) $n^{\frac{1}{6}} \times n^{\frac{1}{6}}$

(c) $n^{\frac{1}{4}} \div n^{\frac{1}{8}}$ (d) $\dfrac{\left(2n^{\frac{1}{2}}\right)^3}{\left(4n\right)^{\frac{1}{2}}}$

3. If $x^y = 5$, find the value of:

(a) x^{2y} (b) $x^{3y} - 5$

(c) x^{-y} (d) x^{y+1}

(e) x^{2y-2} (f) $\left[\left(x^y + 1\right)\left(x^y - 1\right)\right]^{-1}$

Indices and Algebra (1) page 2

4. Solve the following for x.
   (a) $2^{x+1} = 8$ (b) $8^{x-2} = 4^3$

   (c) $\left(\frac{1}{2}\right)^{x+1} = 32$ (d) $81^{2x-1} = \frac{1}{3}$

5. If $x^y = a$ and $x^{2y} - 2 = 14$, find the value of a and list the values of x for $y = \pm 1$ and $y = \pm 2$.

6. Express $X^2 - Y^2$ in terms of numbers and a with positive powers if $X = 3^a + 3^{-a}$ and $Y = 3^a - 3^{-a}$.

Indices and Algebra (2)

1. Express the following in the form x^y.

(a) $2^2 + 2^2$

(b) $3^2 + 3^2 + 3^2$

(c) $4^3 + 4^3 + 4^3 + 4^3$

(d) $2^3 + 2^3 + 2^3 + 2^3$

(e) $4^a + 4^a$

(f) $4 \times 2^a + 4 \times 2^a$

2. Simplify the following.

(a) $\dfrac{2d^{-2k} \times 3^{-4} e^{-2k}}{2^{-3} d^{-k} \times 3^2 e^{5k}}$

(b) $\dfrac{4d^{\frac{1}{2}k} \times 16^{-1} e^{-2\frac{1}{2}k}}{2^{-1} d^{-k} \times 32^2 e^{5\frac{1}{2}k}}$

(c) $\dfrac{2^{2k+2} + 4^{k-3}}{4^{k-1} + 2 \times 2^{2k-3}}$

(d) $\dfrac{3^{2k-2} + 4 \times 3^{2k-3}}{9^{k-1} - 3^{2k-3}}$

Indices and Algebra (2) page 2

3. Express the following with positive powers of a.

(a) $\dfrac{1-a^{-1}}{1+a^{-1}}$

(b) $\dfrac{a^{-2}+1}{a^{-2}-1}$

(c) $\dfrac{2^{a+2}-4}{2^{a}-1}$

(d) $\dfrac{3\times 2^{a}-\left(\frac{1}{3}\right)^{-1}\times 2^{3a-2}}{2^{a}-2\times\left(2^{a-1}\right)^{3}}$

4. Simplify the following.

(a) $\frac{2}{3}\left(ac^{4}h^{2}\right)^{-9}\times\frac{3}{18}\left(a^{-3}c^{4}h\right)^{-5}\div\left(a^{-8}c^{-2}h^{7}\right)^{3}$

(b) $\frac{5}{6}n^{4}\left(c^{8}v^{-6}b^{-8}\right)^{-\frac{1}{2}}\times 36\left(n^{4}vb^{-6}c^{a}\right)^{\frac{1}{4}}$

Solutions

Index Rules and Algebra

1. (a) (i) 1.732050808 (ii) 2.236067977

   (b) It would appear, to nine decimal places that $a^{\frac{1}{2}} = \sqrt{a}$.

   (c) $\left(a^{\frac{1}{2}}\right)^2 = a^{\frac{1}{2}} \times a^{\frac{1}{2}} = a^{\frac{1}{2}+\frac{1}{2}} = a^1 = a$, $\left(\sqrt{a}\right)^2 = \sqrt{a} \times \sqrt{a} = a$, so $a^{\frac{1}{2}} = \sqrt{a}$.

2. (a) (i) $2^0 = 1$ (ii) $(-23)^0 = 1$ (b) (i) 1 (ii) a^0

   (c) $a^0 = 1$

3. (a) (i) $\frac{1}{2}$ (ii) $\frac{1}{4}$ (b) (i) a^{-1} (ii) $\frac{1}{a}$

   (c) $a^{-1} = \frac{1}{a}$, then $a^{-1} = \frac{1}{a}$

4. (a) $5^3 = 5 \times 5^2 = 5 \times 25 = 4 \times 25 + 1 \times 25 = (2 \times 5)^2 + (5)^2 = (10)^2 + (5)^2 = 125$

5.

$$(1+2)^2 - 1^2 = 9 - 1 = 8 = (2)^3$$

$$(1+2+3)^2 - (1+2)^2 = 36 - 9 = 27 = (3)^3$$

$$(1+2+3+4)^2 - (1+2+3)^2 = 100 - 36 = 64 = (4)^3$$

$$(1+2+3+4+5)^2 - (1+2+3+4)^2 = 225 - 100 = 125 = (5)^3$$

$$(1+2+3+.....+n)^2 - (1+2+3+....(n-1))^2 = n^3$$

Indices and Number

2. (a) $\frac{3}{8}$ (b) $\sqrt{2}$
3. (a) 2 (b) 125 (c) 8 (d) 128 (e) 4

   (f) $\frac{1}{11}$ (g) $\frac{1}{243}$ (h) $\frac{1}{25}$
4. (a) 3^6 (b) $2^{(a+3)}$ (c) $3^{(b+3)}$ (d) $5^{(a-1)}$ (e) 2^{3a}

   (f) $2^{(4-a)}$ (g) 2^{2a} (h) $3^{(3a-2b)}$
5. (a) 5 (b) 36 (c) $1\frac{24}{25}$ (d) 16
6. (a) 3125 (b) $2^5 \times 3^{1\frac{7}{12}}$

Indices and Algebra (1)

1. (a) $x^a \times x^b = x^{a+b}$ (b) $\frac{x^a}{x^b} = x^{a-b}$ (c) $\left(x^a\right)^b = x^{ab}$

   (d) $\sqrt[a]{x} = x^{\frac{1}{a}}$ (e) $a^0 = 1$ (f) $x^{-a} = \frac{1}{x^a}$

2. (a) $n^{2\frac{1}{2}}$ (b) $n^{\frac{1}{3}}$ (c) $n^{\frac{1}{8}}$ (d) $4n$
3. (a) 25 (b) 120 (c) $\frac{1}{5}$ (d) $5x$

   (e) $\frac{25}{x^2}$ (f) $\frac{1}{24}$

4. (a) $x = 2$ (b) $x = 4$ (c) $x = -6$ (d) $x = \frac{3}{8}$

5. $a = 4$

x	y
2	2
4	1
$\frac{1}{4}$	-1
$\frac{1}{2}$	-2

6. 4

Indices and Algebra (2)

1. (a) 8 (b) 27 (c) 256 (d) 32
   (e) 2^{2a+1} (f) 2^{a+3}

2. (a) $\dfrac{16}{729d^k e^{7k}}$ (b) $\dfrac{d^{1\frac{1}{2}k}}{2048e^{8k}}$ (c) $8\frac{1}{32}$ (d) $3\frac{1}{2}$

3. (a) $\dfrac{a-1}{a+1}$ (b) $\dfrac{1+a^2}{1-a^2}$ (c) 4 (d) 3

4. (a) $\dfrac{a^{30}}{9c^{50}h^{44}}$ (b) $30n^5v^{3\frac{1}{4}}b^{2\frac{1}{2}}c^{\left(\frac{a}{4}-4\right)}$

Chapter 7

Measurement

Areas of Interest

Perimeter – Straight Sides

Perimeter – Curved Sides (1)

Perimeter – Curved Sides (2)

Area and Perimeter – Straight Sides

Area and Perimeter – Curved Sides

Area – Straight Sides (1)

Area – Curved Sides (2)

Volume

Volume, Surface Area and Capacity

Volume and Surface Area

Perimeter – Straight Sides

1. Find the perimeter of these shapes.

(a)

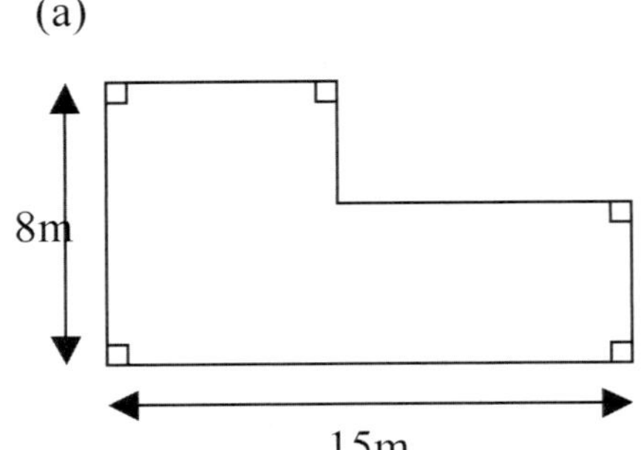

(b)

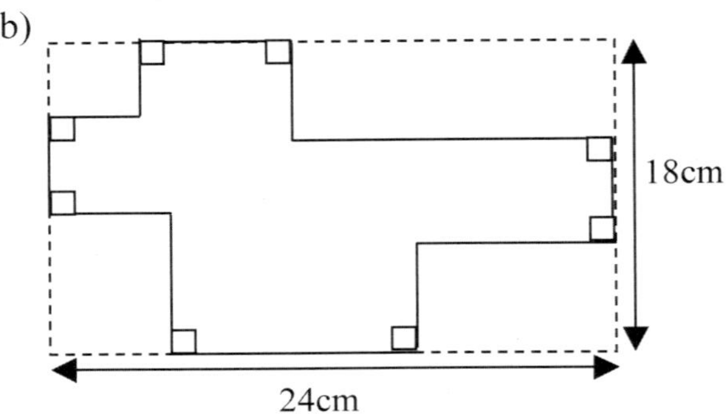

2. Find the perimeter of the shapes made with congruent parallelograms.

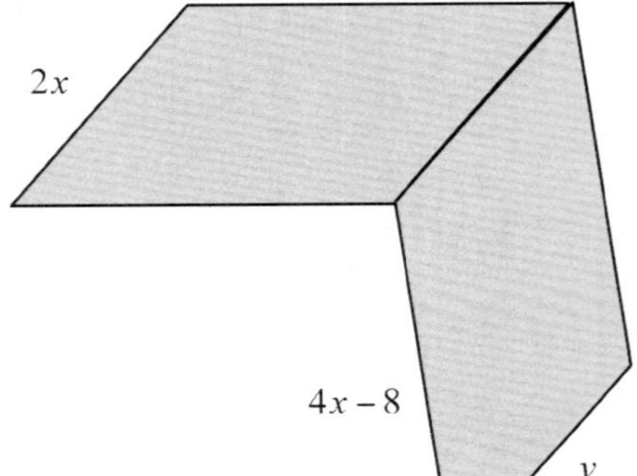

(b)

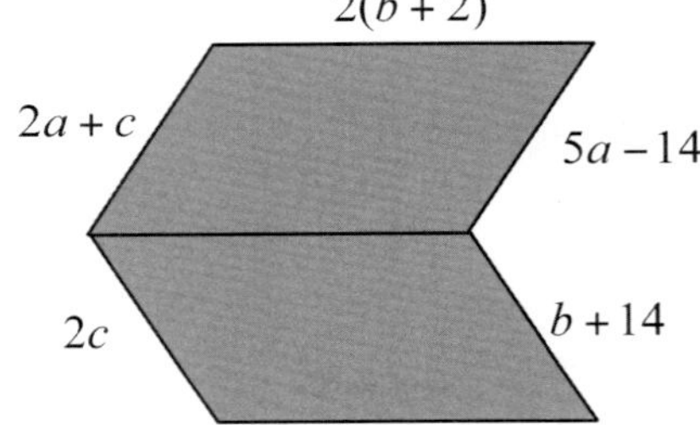

Perimeter – Straight Sides page 2

3. A rectangle is cut into two shapes.
   (a) Find the lengths show as a and b.

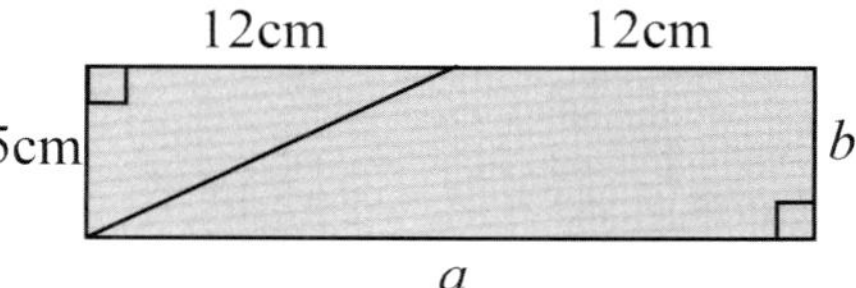

   (b) The shapes are joined to make a parallelogram and a right-angled triangle. Find the perimeter of the:
      (i) parallelogram

      (ii) right-angled triangle.

4. A square of edge length 36cm is divided into four identical smaller squares by drawing a cross over it.
   (a) Find the perimeter of the white cross if the sum of the perimeters of the smaller squares is 160cm.

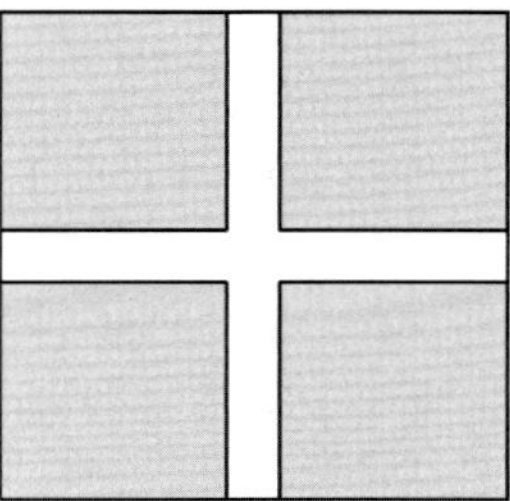

   (b) Find the perimeter of the cross if the perimeter of the smaller squares is twice that of the perimeter of the original square.

Perimeter – Curved Sides (1)

1. The minute hand of a clock is 10cm in length.
   (a) Find the distance in exact form that the end of the minute hand travels in:
   (i) 20 seconds (ii) 20 minutes (iii) one day.

   (b) The hour hand of the clock is 5cm in length. How much further does the end of the minute hand travel in an hour compared to the distance travelled by the end of the hour hand?

2. A coin of radius 2cm is rolled around a square, an equilateral triangle and a regular hexagon each with side lengths 10cm.
   (a) Find the distance travelled by the centre of the coin after it has travelled once around the:
   (i) square (ii) equilateral triangle (iii) regular hexagon.

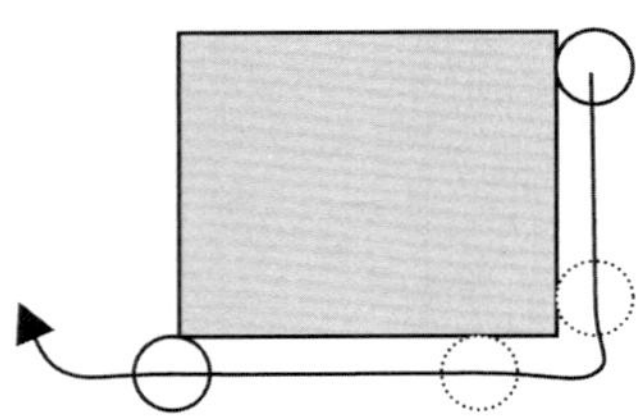
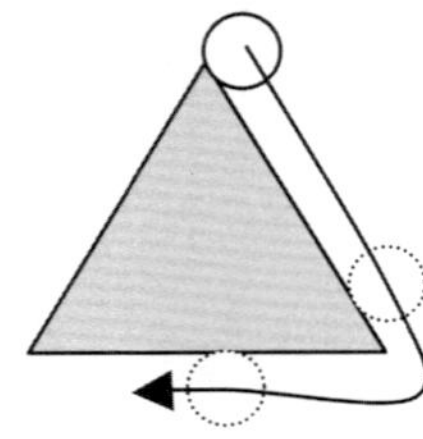
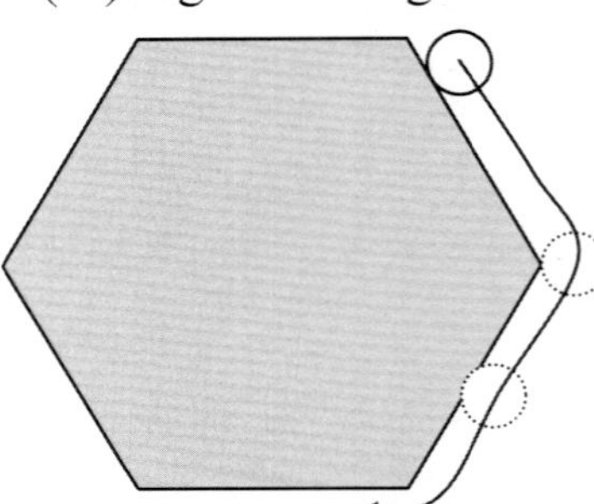

   (b) Find the distance travelled by the centre of a coin with radius 2cm when rolled around a regular n-sided shape, edge length L.

Perimeter – Curved Sides (1) page 2

2. (c) (i) Find the perimeter of (i) a square (ii) an equilateral triangle and (iii) a regular hexagon if a coin with radius 2cm makes 37 revolutions whilst travelling around each shape.

(d) A coin with radius of 2cm is rolled around an n-sided shape. Express the perimeter of the shape if the coin makes r revolutions whilst travelling around the shape.

3. The ratio of the front to the back wheel of William's bicycle is 2:3. Over 360m the front wheel makes 160 more revolutions than the back wheel.

(a) Find the number of revolutions, n, that the front wheel makes.

(b) Find the difference between the circumference of the front and back wheel.

(c) Find the radii of the front and back wheels.

Perimeter – Curved Sides (2)

1. Find the perimeter of the following shapes in terms of a.

(a) (b) (c)

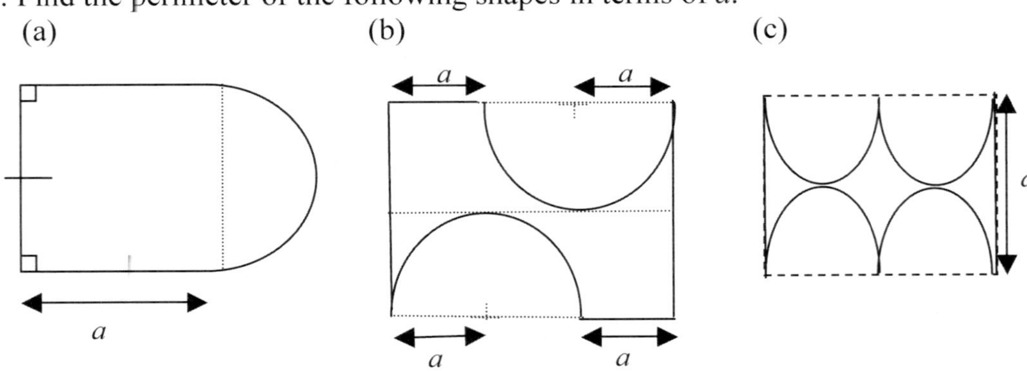

2. Two squares have four equal circular sections removed with lengths as shown to make two shaded shapes. Square 1 has quarter circles removed while square 2 has semicircles removed.

(a) For each shape, find the perimeter of the shaded shapes in terms of the pronumerals.

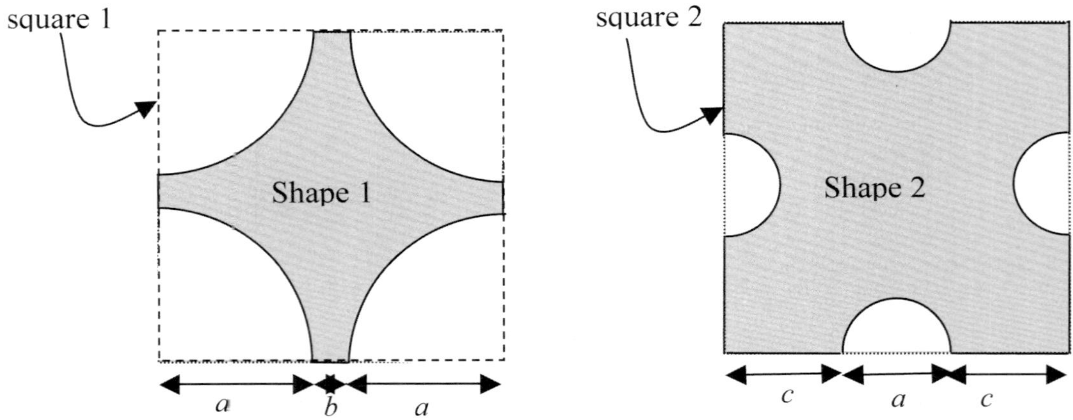

(b) (i) Find the relationship between b and c so that the perimeters of the shaded shapes are the same.

Perimeter – Curved Sides (2) page 2

2. (b) (ii) Find the size of the square 1 in terms of a and c, and find the difference between the edge lengths of the squares.

3. Two identical circles both with radii of b units have a tight rubber band stretched around them. They are a units apart as shown.

   (a) Find the total length of the rubber band in terms of a and b.

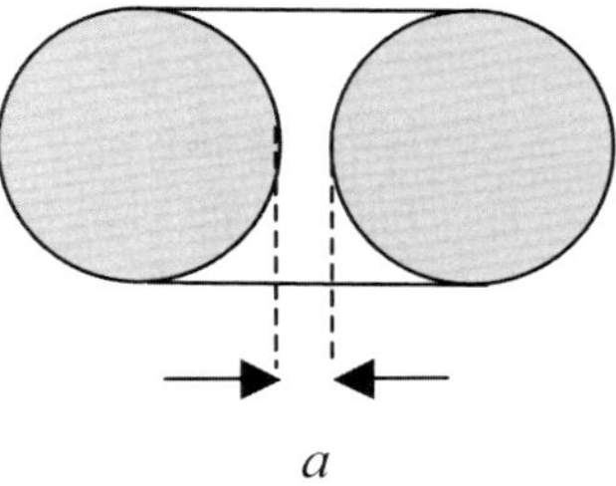

   (b) Find an expression for a if the length of the rubber band is:

   (i) $2b(\pi + 3)$ (ii) $4b(\pi + 1)$

   (c) Three identical circles with radii of b units have a tight rubber band stretched over them. They are a units apart as shown.

   (i) Find the length of the rubber band in terms of a and b.

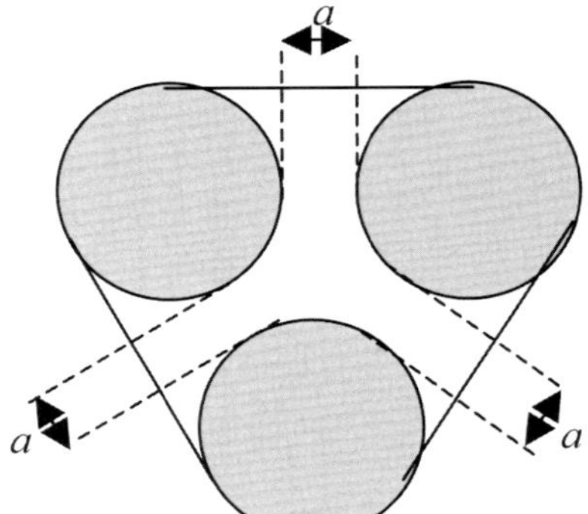

   (ii) Find an expression for a if the rubber band is $2b(\pi + 3)$.

Area and Perimeter – Straight Sides

1. A shaded hexagon is shown on centimetre grid paper.
   (a) (i) Draw a triangle, a parallelogram and a trapezium on the grid so that they have the same perimeter as the hexagon.

   (ii) Use Pythagoras' Theorem to show that the perimeter of each shape is $\left(4\sqrt{5}+8\right)$cm .

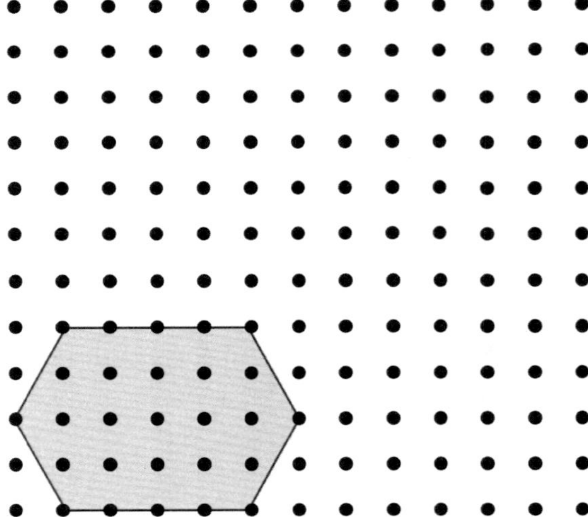

   (b) Find the area of each shape and hence, find the ratio of the areas of the shapes:

   (i) hexagon to triangle

   (ii) hexagon to trapezium

   (iii) hexagon to parallelogram

   (iv) triangle to trapezium

   (v) triangle to parallelogram

   (vi) trapezium to parallelogram.

2. The rectangle has length a and width b.
   (a) If the area of the rectangle is the same as its perimeter, write an equation expressing b in terms of a.

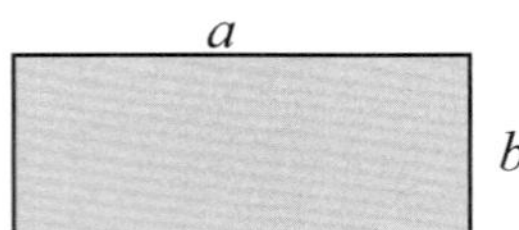

Area and Perimeter – Straight Sides page 2

2. (b) Use the equation above to find the width of the rectangles for the given lengths and show that the perimeter is the same as the area for each rectangle.
   (i) length 10cm (ii) length 25cm

   (c) Use algebra to find the width and the length of a rectangle whose length is twice its width and its area is the same as its perimeter. Verify the solution.

3. The shape is made up with long lengths and short lengths. The longer lengths are twice the length of the shorter lengths. If the area of the shape is 1 575cm^2, find the perimeter of the shape.

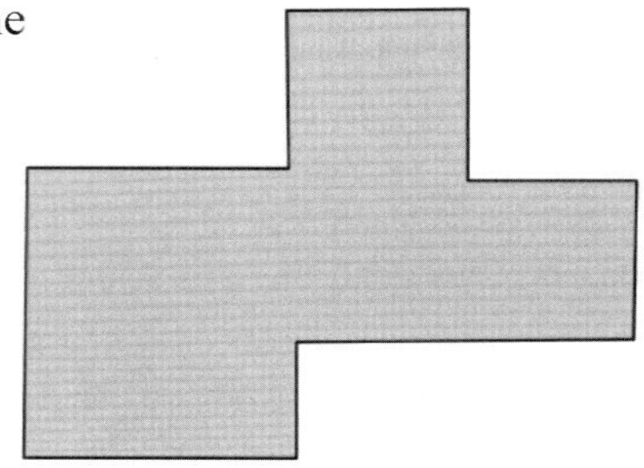

Area and Perimeter – Curved Sides

1. (a) The semicircle has its area the same as its perimeter.
   (i) Find the radius and diameter of the semicircle in terms of π .

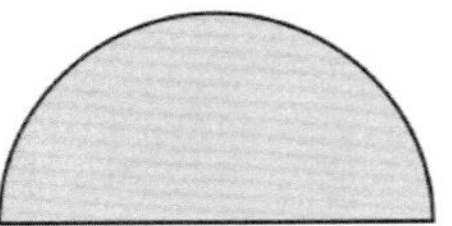

   (ii) If the radius was doubled, find the new perimeter and area. Hence, find how many times bigger the ratio of the perimeter to the area of the larger semicircle is compared to the ratio of the perimeter to the area of the original semicircle.

   (b) (i) When the area is quartered, find the ratio of the new perimeter to the old perimeter.

   (ii) When the perimeter is doubled, find the ratio of the new area to the original area of the semicircle.

2. Four identical circles each with radii, r are arranged as shown.
   (a) Show that $\theta = 45°$.

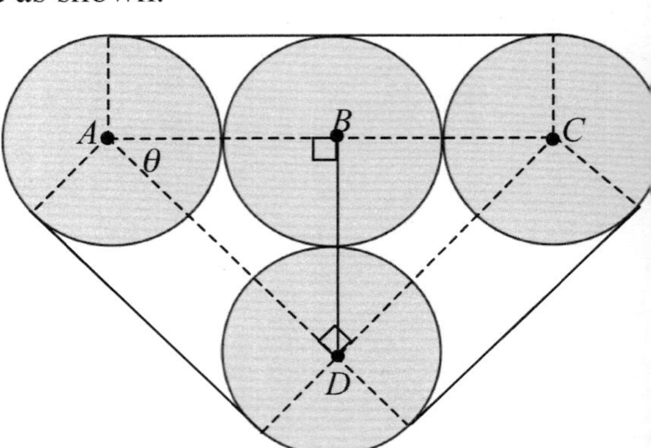

   (b) A tight rubber band is stretched around the circles.
   (i) Find the length of the rubber band expressed in exact form in terms of r.

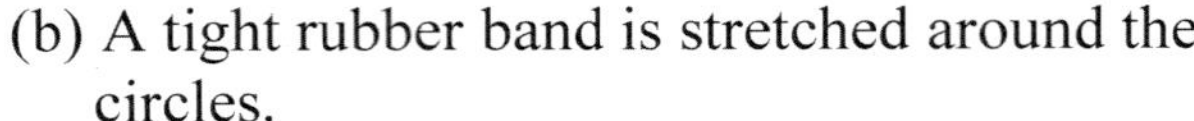

2. (b) (ii) Find the total area of the shape inside the rubber band in terms of r.

Area and Perimeter – Curved Sides page 2

3. Five concentric circles result in the areas A, B, C, D and E being shown on the diagram.

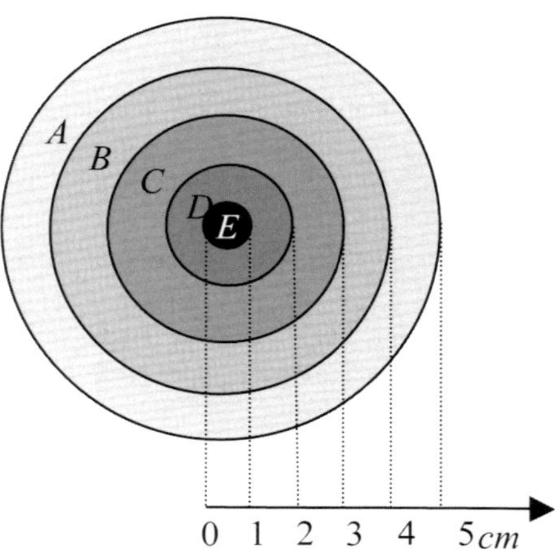

(a) Find the ratio of area E to:

(i) D (ii) C (iii) B (iv) A

(b) Find the ratio of the:

(i) circumference of the first circle to the area of E

(ii) circumference of the outermost circle to the sum of the areas of regions B, C and D

(iii) circumference of the middle circle to the area of the regions outside that circle.

Area – Straight Sides (1)

1. Assuming that the area of a rectangle with length L and width W is LW, prove the area formulae for the following.

(a)

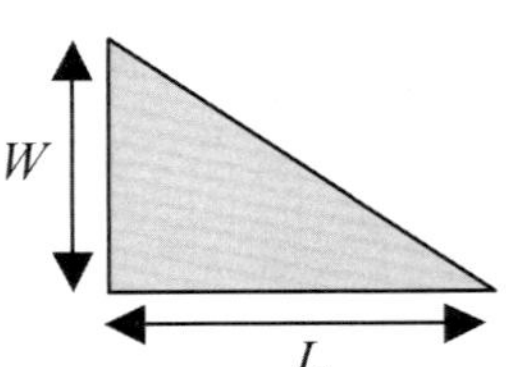

Triangle: $A = \frac{1}{2}LW$

(b)

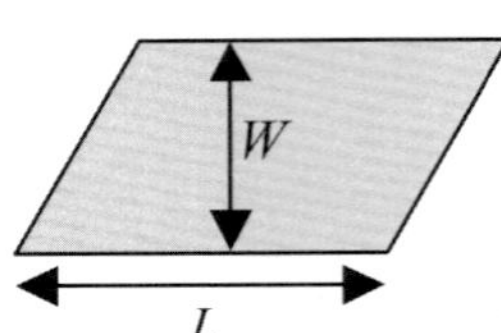

Parallelogram: $A = LW$

(c)

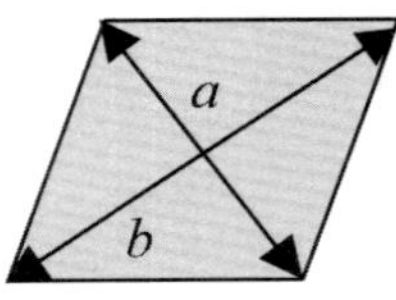

Rhombus: $A = \frac{1}{2}ab$

2. In the diagram, $AB = 8\text{cm}$, $BC = 6\text{cm}$ and $BC = DE$ and the area of ABD is 40cm^2.

(a) Find the length CD and hence, find the area of triangle ACD.

(b) Find the areas of BCF and CDF. Hence, find the ratio of the areas of those triangles.

Area – Straight Sides (1) page 2

3. Three shapes are shown on centimetre grid paper.
   (a) Find the area of the parts of each of the shapes, which are:
      (i) single shaded

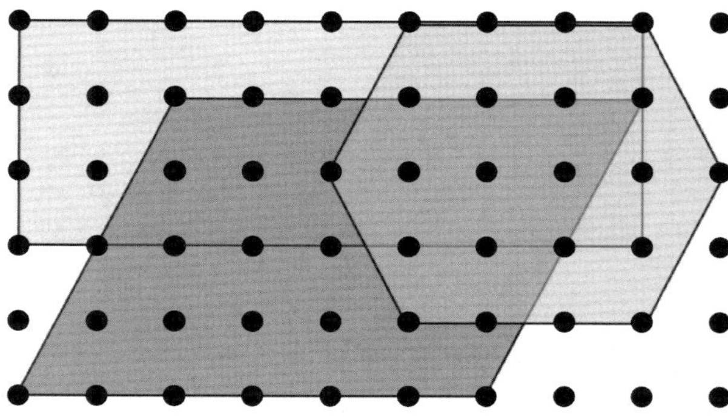

      (ii) double shaded

      (iii) triple shaded.

   (b) Find the proportion of the parallelogram, which is:
      (i) double shaded (ii) triple shaded.

   (c) Find the proportion of the rectangle, which is:
      (i) double shaded (ii) triple shaded.

   (d) Find the proportion of the hexagon, which is:
      (i) double shaded (ii) triple shaded.

Area – Straight Sides (2)

1. A gardener is employed to lay a path of width 1.2m around a $8\text{m} \times 4.5\text{m}$ rectangular swimming pool.
   (a) Find the area of the paved path.

   (b) Find the dimensions of each brick if there are 50 bricks per square metre and each side is a multiple of ten.

   (c) Find the number of paving bricks that should be ordered if there is a 5% breakage rate for the job showing the pattern that minimizes the time to complete the job.

2. Pieces of cardboard of size 7cm by 12cm are to be cut from a sheet 2m by 1.6m.
   (a) Find the maximum number of pieces that can be cut from the original sheet.

   (b) (i) Find the area that is wasted.

   (ii) If the waste is collected and glued together to make another sheet, find the maximum number of pieces that can be cut from it along with the area left over.

Area – Straight Sides (2) page 2

3. The diagram shows a square with side length c, and two triangles ABE and ADE.

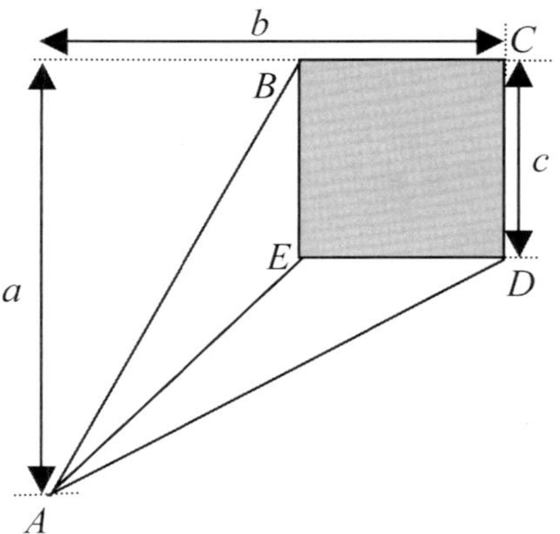

(a) Find the area of the triangles in terms of a, b and c.

(i) ABE

(ii) ADE

(b) Find the ratio of the areas of the triangles.

(c) Find the area of the shape $ABCD$.

Area – Curved Sides (1)

1. The length OA, shown for the circular sections is a radius of r units. Express the ratio of the shaded areas to the total area of the shapes in exact form.

(a)

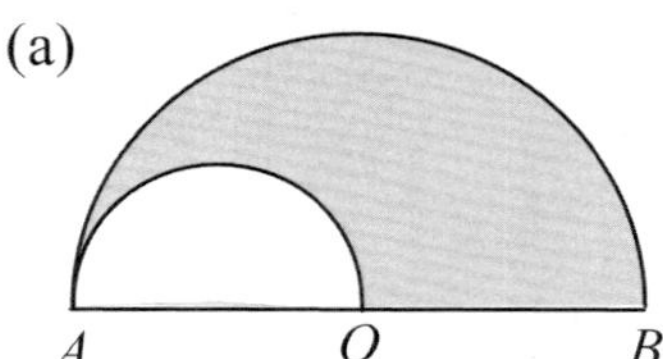

(b)

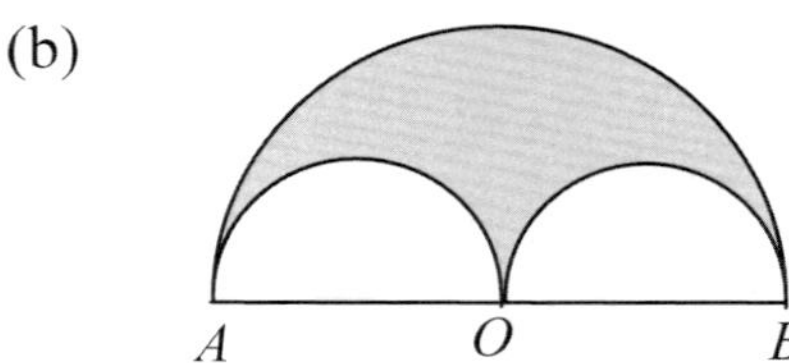

(c)

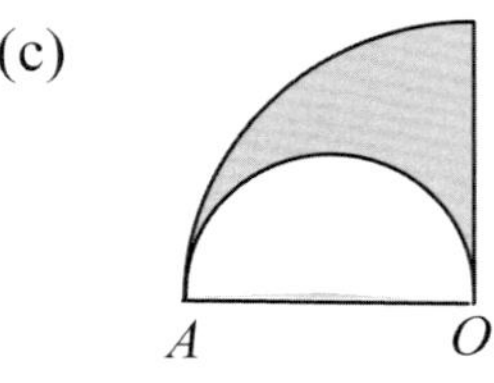

(d)

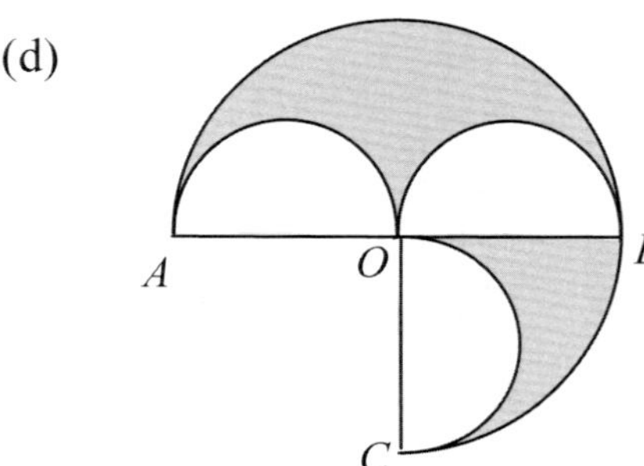

2. The semicircle has a triangle ABC placed inside it so that AB is the diameter of the circle of length 10cm .
   (a) State the angle $\angle ACB$.

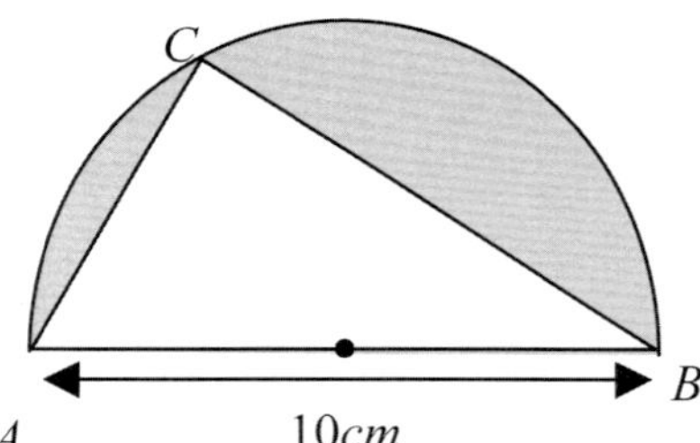

   (b) If $AC = 6\text{cm}$, find BC using Pythagoras' theorem.

   (c) Find the shaded area expressed in exact form.

Area – Curved Sides (1) page 2

3. A goat is tethered in the field shown where $\angle BAC = 36.9°$

 (a) Find the area over which it can graze if it is tethered to the point A with a rope of length:

 (i) 10 metres

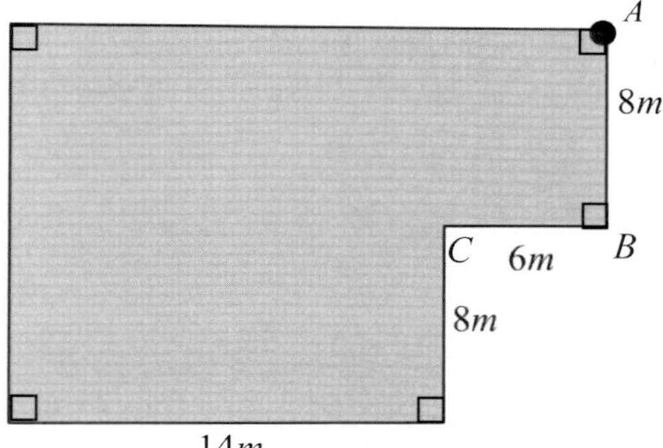

 (ii) 20 metres.

 (b) Find the ratio of the area that the goat cannot reach when tethered by a 10 metre rope compared to a 20 metre length of rope.

Area – Curved Sides (2)

1. Four identical circles each with radius r, are placed so that each circle touches two others. The centres of the circles are the vertices of a square and three shaded areas are shown light grey, dark grey and stripes.

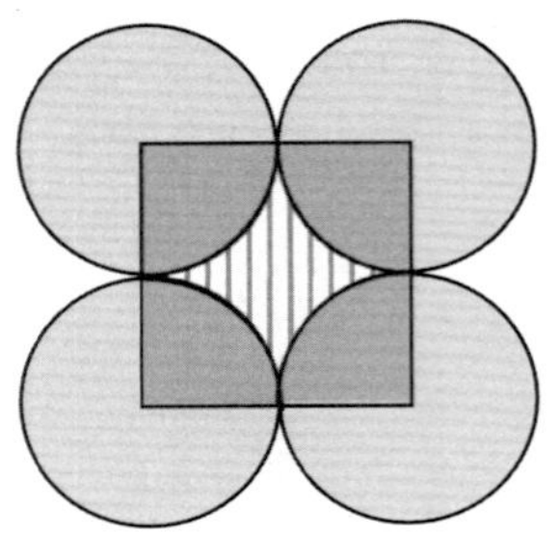

   (a) Find the ratio of:

   (i) the light grey area to the dark grey area

   (ii) the striped area to the dark grey area.

   (b) Find the area of the circle that touches the:

   (i) inside of the four circles

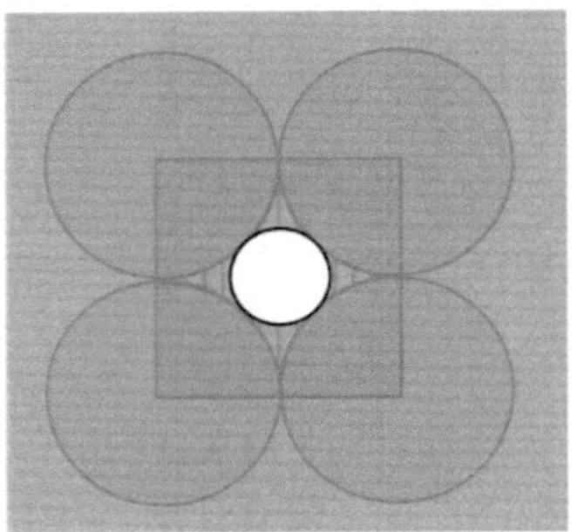

   (ii) outside of the circles.

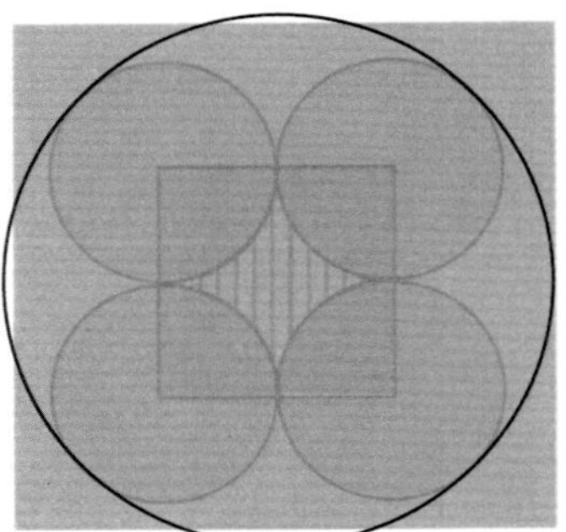

2. Two semicircles are drawn above the line AB and four semicircles are drawn below the line AB.

   (a) Find the ratio of the area contained by the semicircles above the line AB compared to the area of the semicircles below the line AB.

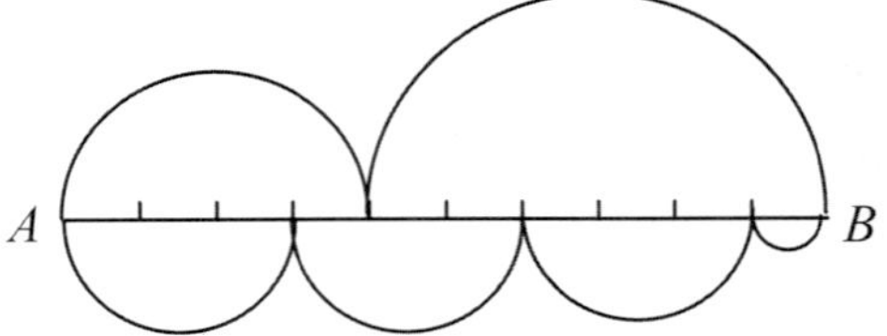

   (b) Express the ratio of the areas of the semicircles in descending order.

Area – Curved Sides (2) page 2

2. (c) Find the radius of a circle, which would have the same area as the collection of semicircles:
   (i) above the line AB (ii) below the line AB.

3. Seven identical circles with radii r, are packed inside a larger circle.
   (a) Find the area of the shaded shape in terms of r.

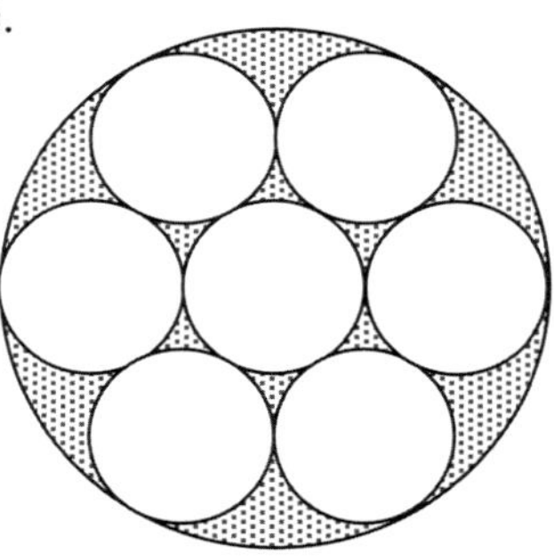

   (b) Find the ratio of the area of the larger circle to the sum of the areas of the smaller circles.

   (c) A square is placed around the circles as shown. Find the unshaded area in the diagram.

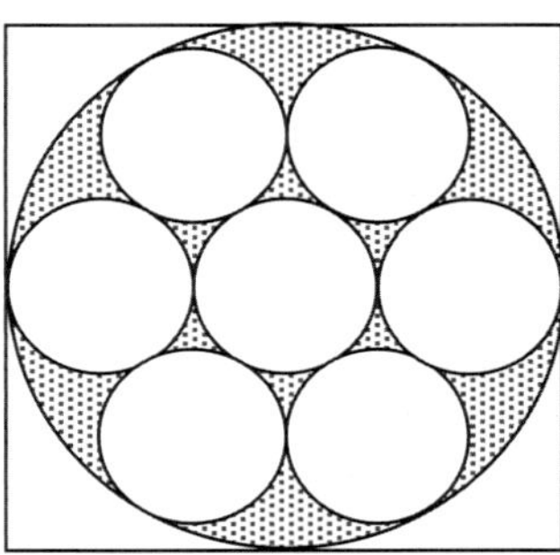

Volume

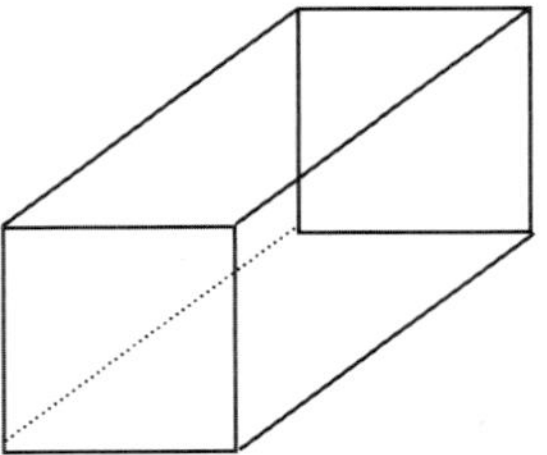

1. A cuboid with a square end (width and height) has a length greater than either dimension of its square face.
   (a) Find the dimensions of the cuboids with integer value lengths so that the volume of the cuboids are 128cm^3

   (b) (i) If the end dimensions (width and height) are halved, find the length of each cuboid for the volume to remain unchanged.

   (ii) If the end dimensions (width and height) are doubled, find the length of each cuboid for the volume to remain unchanged.

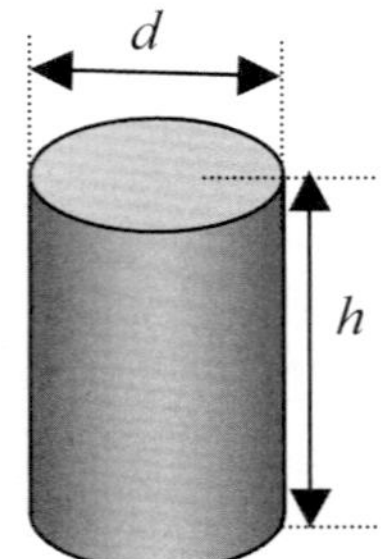

2. A cylinder with height, h and diameter, d is shown.
   (a) (i) Express the volume of the cylinder in terms of h and d.

   (ii) Express the volume of the cylinder in terms of d where the height is twice the diameter.

   (iii) Express the volume of the cylinder in terms of d where the diameter is n-times its height.

   (b) Find the volume of a cylinder with height $(a + 2)$ and a diameter twice that of its height, in fully expanded form.

Volume page 2

3. A 12cm cube has a cylinder radius, r and depth, D drilled through the front face.

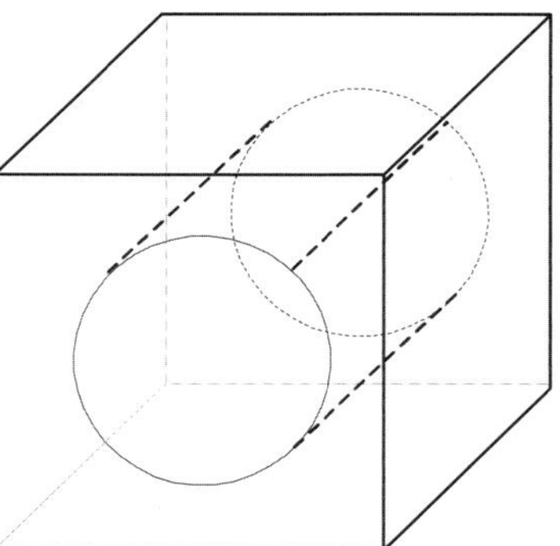

(a) Express r in terms of D if three-quarters of the cube remains.

(b) (i) If three-quarters of the cube remains, find the radius of the cylinder when the hole is drilled all the way through the cube.

(ii) Between what limits can the radius of the cylinder range? Find the depth of the cylinder with maximum radius.

(c) Find the percentage of the cube that remains after the largest cylinder has been removed.

Volume, Surface Area and Capacity

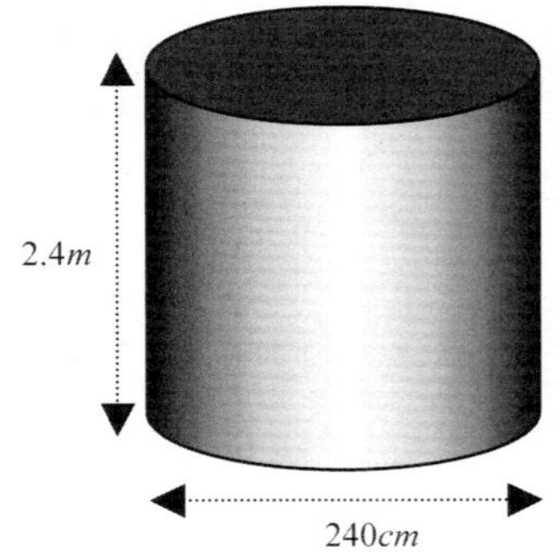

1. A cylindrical water tank with dimensions shown is completely filled after a wet winter.
   (a) Find the volume of the tank in:
      (i) cubic metres

      (ii) cubic centimetres.

   (b)(i) Unfortunately the tank is punctured during target practice prior to the hunting season and half of the water leaks out before it is fixed. If the water leaked out at a constant rate of 2 litres per second, find the number of minutes to the nearest second for this to happen.

      (ii) If one litre of water weighs one kilogram then what weight of water remains in the tank?

   (c) A white rectangular stripe is painted around the tank 70cm wide and is labelled as shown.

      (i) Find the surface area of the stripe to the closest square centimetre.

      (ii) What percentage of the wall of the tank does the stripe occupy to the nearest whole number?

   (d) The remainder of the water is poured from the tank into a cube, which it fills completely.
   Find the dimensions of the cube in centimetres to one decimal place.

Volume, Surface Area and Capacity page 2

2. This is a plan of a new petrol tank to be placed under a petrol station where all sides are at right-angles.

   (a) Find the volume in litres when the tank is filled to the top.

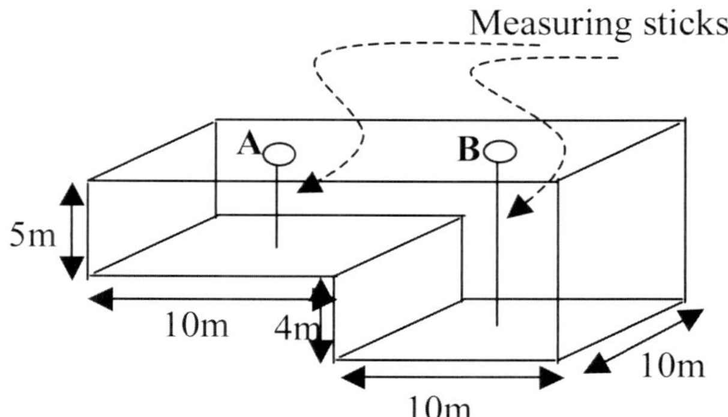

Observation holes are placed at points **A** and **B** shown on the diagram where measuring sticks can be inserted to measure the height of the petrol from the bottom of the two sections of the tank.

   (b) If one million litres is placed into the tank, find the height of the petrol above the bottom of the tank at observation points:

   (i) **A** (ii) **B**

Volume and Surface Area

1. A right triangular-based prism with depth 10cm with edge lengths 6cm, 8cm and 10cm is shown. A smaller block half the size of the bigger block is stuck on the sloping face so that its height is less that its depth.

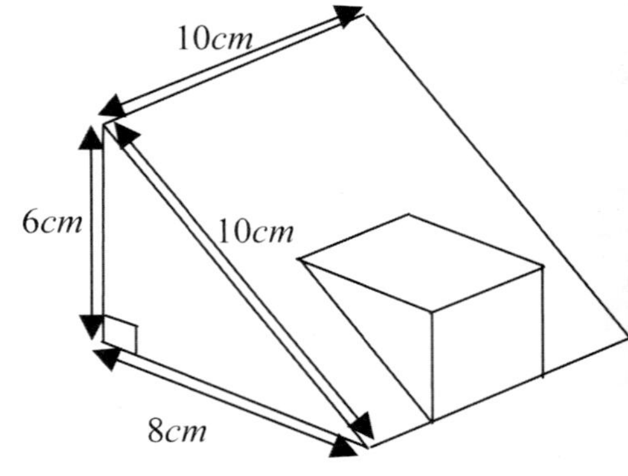

(a) Find the volume of the solid.

(b) Find the surface area of the solid.

2. Wedges of cheese are to be wrapped in silver paper.
   (a) Find the volume and total surface area of the wedge to the nearest whole number.

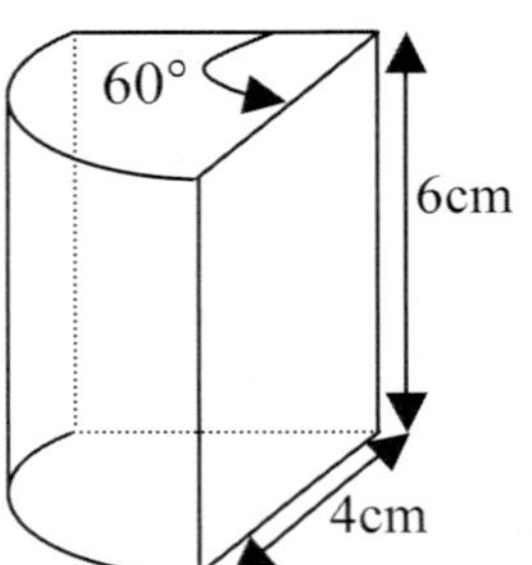

Liquid cheese is poured in to the container at the rate of $1\frac{1}{2}$ milliliters per second.

(b) Find the height of the water after:
(i) 15 seconds (ii) half a minute.

Volume and Surface Area page 2

3. A cube with edge length 12cm has square-ended tunnels cut through each face, right through the tunnel. The squares on the faces of the cube are 6cm.
   (a) (i) Find the volume of the solid.

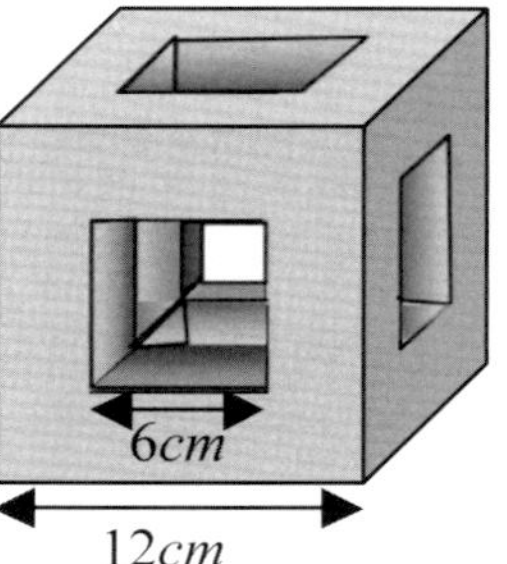

   (ii) Find the surface area of the solid.

(b) When a cube with edge length b cm has tunnels cut in it so that the edge length of the square a cm
   (i) Find the volume of the solid in terms of a and b.

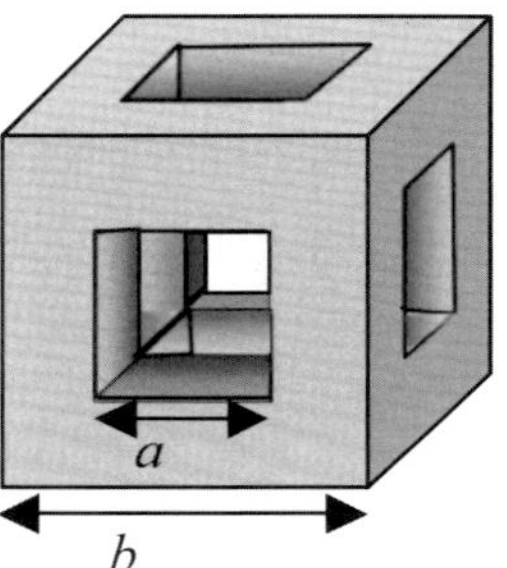

   (ii) Find the surface area of the solid in terms of a and b.

Solutions

Perimeter – Straight Sides

1. (a) 46m (b) 84m
2. (a) 168cm (b) 400
3. (a) $a = 24,\ b = 5$ (b) (i) 74cm (ii) 60cm
4. (a) 144cm (b) Width of the white cross: 27cm, Perimeter of White cross: 144cm

Perimeter – Curved Sides (1)

1. (a) (i) $= \frac{\pi}{9}$ cm (ii) $6\frac{2}{3}\pi$ cm (iii) 480π cm
   (b) $19\frac{1}{6}\pi$ cm
2. (a) (i) $(40 + 4\pi)$cm (ii) $(30 + 4\pi)$cm (iii) $(60 + 4\pi)$cm
   (b) $(Ln + 4\pi)$
   (c) (i) 144π cm (ii) 144π cm (iii) 144π cm
   (d) $4\pi(r - 1)$cm
3. (a) 480 revs (b) $\frac{3}{8}$m
   (c) Radius of front wheel: $\dfrac{3}{8\pi}$m , radius of back wheel: $\dfrac{9}{16\pi}$m

Perimeter – Curved Sides (2)

1. (a) $a(3 + \frac{1}{2}\pi)$ (b) $2a(3 + \pi)$ (c) $2a(1 + \pi)$
2. (a) Shape 1: $P = 4b + 2a\pi$, Shape 2: $P = 8c + 2a\pi$
   (b) (i) $4b + 2a\pi = 8c + 2a\pi,\ \therefore b = 2c$
   (ii) Square 1: $2a + 2c$, square 2: $a + 2c$. Square 1 is a units longer than square 2.
3. (a) $2b\pi + 2a + 4b$
   (b) (i) $a = b$ (ii) $a = b\pi$
   (c) (i) $6b + 3a + 2b\pi$ (ii) $a = 0$

Area and Perimeter – Straight Sides

1. (b) (i) $20:8 = 5:2$ (ii) $20:16 = 5:4$ (iii) $20:16 = 5:4$
   (iv) $8:16 = 1:2$ (v) $8:16 = 1:2$ (vi) $16:16 = 1:1$
2. (a) $b = \dfrac{2a}{(a - 2)}$
   (b) (i) 25cm^2 (ii) $54\frac{8}{23}\text{cm}^2$
   (c) length 6 and width 3
3. 210cm

Area and Perimeter – Curved Sides

1. (a) (i) $r = \dfrac{2(\pi + 2)}{\pi}$, $d = \dfrac{4(\pi + 2)}{\pi}$ (ii) $P = 2r(\pi + 2)$, $A = 2\pi r^2$. 2:1
   (b) (i) 2:1 (ii) 1:4
2. (b) (i) $2r(2 + \pi + 2\sqrt{2})$ (ii) $r^2(\pi + 8 + 4\sqrt{2})$
3. (a) (i) $1:8$ (ii) $1:16$ (iii) $1:24$ (iv) $1:32$
   (b) (i) 4:1 (ii) 3:4 (iii) 5:14

Area – Straight Sides (1)

2. (a) $CD = 10\text{cm}$, 70cm^2 (b) Area 15cm^2, Area 15cm^2, Ratio of the areas: $1:1$
3. (a) (i) Rectangle: $7\frac{3}{4}$, Parallelogram: 10, Hexagon: $3\frac{1}{4}$

   (ii) Hexagon and rectangle: $4\frac{1}{4}$, Hexagon and parallelogram: 2,
   Rectangle and parallelogram: $5\frac{1}{2}$

   (iii) Area of rectangle: 24, parallelogram: 24, hexagon: 16

   (b) (i) $24:(2+5\frac{1}{2}) = 24:7\frac{1}{2} = 16:5$ (ii) $24:6\frac{1}{2} = 48:13$

   (c) (i) $24:9\frac{3}{4} = 32:13$ (ii) $24:6\frac{1}{2} = 48:13$

   (d) (i) $16:6\frac{1}{4} = 64:25$ (ii) $16:6\frac{1}{2} = 32:13$

Area – Straight Sides (2)

1. (a) 35.76m^2 (b) $10\text{cm} \times 20\text{cm}$ (c) 1 788, 1 882
2. (a) 364 pieces (b) (i) 1 424cm^2 (ii) 16 pieces, 80cm^2
3. (a) (i) $\frac{1}{2}c(b-c)$ (ii) $\frac{1}{2}c(a-c)$

   (b) $(b-c):(a-c)$

   (c) $\frac{1}{2}c(a+b)$

Area – Curved Sides (1)

1. (a) $\frac{3}{8}\pi r^2$ (b) $\frac{1}{4}\pi r^2$ (c) $\frac{1}{8}\pi r^2$ (d) $\frac{3}{8}\pi r^2$
2. (a) 90° (b) 8cm (c) $\left(12\frac{1}{2}\pi - 24\right)\text{cm}^2$
3. (a) (i) 56.2m^2 (ii) 152.8m^2 (b) $1.8:1$

Area – Curved Sides (2)

1. (a) (i) 3:1 (ii) $(4-\pi):\pi$

   (b) (i) $\pi r^2\left(3-\sqrt{2}\right)$ (ii) $\pi r^2\left(3+\sqrt{2}\right)$

2. (a) 13:7 (b) $36:16:9:1$ (c) (i) $r = \sqrt{13}$ (ii) $r = \sqrt{7}$
3. (a) $2\pi r^2$ (b) 9:7 (c) $2r^2(18-\pi)$

Volume

1. (a)

x	y
1	128
2	32
4	8
8	2

1. (b) (i)

x	y
$1 \to \frac{1}{2}$	$128 \to 512$
$2 \to 1$	$32 \to 128$
$4 \to 2$	$8 \to 32$
$8 \to 4$	$2 \to 8$

(ii)

x	y
$1 \to 2$	$128 \to 32$
$2 \to 4$	$32 \to 8$
$4 \to 8$	$8 \to 2$
$8 \to 16$	$2 \to \frac{1}{2}$

2. (a) (i) $V = \frac{1}{4}\pi d^2 h$
   (b) (i) $V = \frac{1}{2}\pi d^3$ (ii) $V = \frac{1}{4}n\pi d^3$
3. (a) (i) $r = 12\sqrt{\frac{3}{\pi D}}$
   (b) (i) r 3.39cm (ii) r: $[3.92, 6)$, D 3.82cm
   (c) $25(4-\pi)$

Volume, Surface Area and Capacity

1. (a) (i) 10.857344m^3 (ii) 10 857.34 litres
   (b) (i) 45 min 14 seconds (ii) Approximately $5\frac{1}{2}$ tonnes
   (c) (i) $52\ 779\text{cm}^2$ (ii) $29\frac{1}{6}\%$
   (d) 175.8cm
2. (a) 1 400 000L
   (b) (i) 3 metres (ii) 7 metres

Volume and Surface Area

1. (a) 270cm^3 (b) 310cm^2
2. (a) $V = 50\text{cm}^3$, $TSA = \left(48 + 10\frac{2}{3}\pi\right)\text{cm}^2$
   (b) (i) 2.69cm (ii) 5.37cm
3. (a) (i) 864cm^3 (ii) $1\ 080\text{cm}^2$
   (b) (i) $b^3 - 3a^2b + 2a^3$ (ii) $6(b+3a)(b-a)$

Chapter 8

Statistics

Areas of Interest

Averages

Statistical Measures

Using Diagrams

Averages

1. In a set of seven numbers, the average of the first three numbers is 12 and the average of the last four numbers is 16. Find the average of the seven numbers.

2. The average of ten numbers was 21. When a number is removed, the average is 22. What number was removed from the original group?

3. In a group of boys and girls, the average age of the boys is 18 and the average age of the girls is 12. Find the ratio of the number of boys to girls if the average of the group is 16.

4. The average score on a test for a group of six students is 38. One student's mark had been recorded as 24 instead of the correct score of 42. Find the correct average for the six students.

5. Four numbers are written in ascending order. If the average of the first and second numbers is 50, the average of the first and third numbers is 60 and the average of the third and fourth numbers is 80, find the average of the second and fourth numbers.

6. Six maths tests each out of 100 are completed each semester. A student has achieved an average of 78 for the first four tests. Find the largest whole percentage average that the student can achieve for the semester.

7. The average of a set of numbers is 56. When two more numbers are included in the set the average is 60. Find the average of the two new numbers if the original set contained 28 values.

Statistical Measures

1. A set of scores has:
   (a) 4 added to each score (b) each score doubled

   (c) each score increased by 50%.

   Find the effect on the (i) mean (ii) median (iii) mode and (iv) range for each of the changes to the scores above.

2. Six values in a data set are {3, 4, 4, 6, 6, 7}. If another value is added to the set, find its value/s if for the new set:
   (a) the mean is the same as the median (b) the median is larger than the mean

   (c) the mean is the same as the range (d) the median is the same as the mode.

3. A data set is {5, 5, 6, 6, 6, 6, 6}.
   (a) Find the next value to be added to the set so that:
      (i) the mean, median and mode are the same

      (ii) the range is less than the mean

      (iii) the range is larger than the median.

Statistical Measures page 2

3. (b) (i) Use algebra to show that the value added to the data set so that the difference between the mean and the range is the same as the median is $4\frac{4}{7}$.

(ii) Verify the solution to the equation and explain why this value does not satisfy the above conditions.

4. There are 12 houses in a short street. The numbers can only be used once.

(a) List the house numbers that can be used so that the mean value of the set is:

(i) 11 (ii) 4

(b) For two house numbers h_1 and h_2, find the value of k so that when $h_1 = a$ and $h_2 = ka$, $a \in Z^+$ the mean of h_1 and h_2 is the same as the range and state the values that a can take.

Using Diagrams

1. These are the scores achieved by students from two classes on a common test.

   Form 7X: 13, 27, 32, 29, 7, 23, 25, 38, 19, 26, 33, 18,
   19, 19, 13, 14, 4, 39, 40, 12, 28, 7, 24

   Form 7Y: 31, 7, 20, 13, 20, 39, 18, 15, 8, 15, 15, 19,
   17, 25, 8, 17, 13, 23, 12, 22, 31, 8, 29

   (a) Set up a back-to-back stem-and-leaf plot to compare the performances of these two Year-7 classes on a recent test.

   (b) Use the ordered stem-and-leaf plot to draw a box-and-whisker plot showing both classes on the same set of axes.

   (c) Use either plot to determine which class performed better on the test. Explain your choice.

Using Diagrams page 2

2. A group of plants was measured after having been exposed to different amounts of light.

Their heights, in centimetres, are as follows:

12.50	6.30	9.80	5.20	12.03
2.01	5.69	9.25	8.95	12.30
14.60	15.20	8.20	9.08	4.12
7.89	11.30	7.89	2.56	10.30
4.90	2.90	2.36	2.01	4.99
2.36	9.06	12.30	0.30	10.10

(a) Set up tally table using intervals of 2cm.

(b) Construct a histogram of the information.

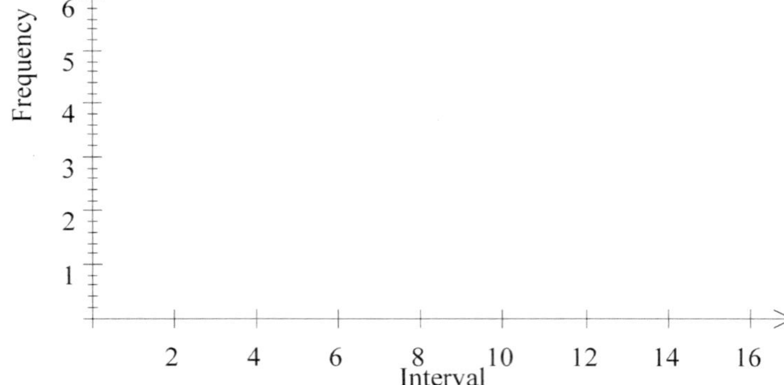

Using Diagrams page 3

2. (c) Construct a pie graph of the information.

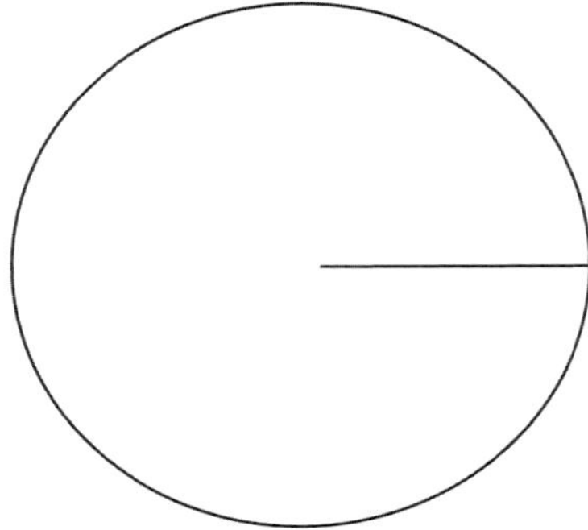

(d) How many plants were at least 5cm tall?

(e) How many plants were at least 10cm tall?

(f) What percentage of plants were at least 9cm tall?

(g) What percentage of plants were less than 9cm tall?

(h) A similar seed that was planted outside in the garden grew to 5.3cm in the same time period as the seeds in the trial. What conclusions can you reach about the effectiveness of the lighting used in the trial?

Solutions

Averages

1. $14\frac{2}{7}$ 2. 12 3. 2:1 4. 41 5. 70 6. 85%
7. 116

Statistical Measures

1. (a) (i) (ii) (iii) the mean, median and mode will all be increased by 4. (iv) the range will not change.
   (b) (i) (ii) (iii) (iv) the mean, median, mode and range will all be doubled.
   (c) (i) (ii) (iii) (iv) the mean, median, mode and range will all be increased by 50%.
2. (a) -2, 5, 12 (b) $[6,11]$, $(-\infty,-3]$ (c) $5\frac{1}{2}$ (d) 4, 6
3. (a) (i) 8 (ii) $[1,11]$ (iii) $[11,\infty)$
4. (a) (i) $\{11\}$, $\{10, 11, 12\}$ (b) $k = 3$, $a \in [1,4]$

Using Diagrams

1. (a)

Form 7X		Form 7Y
4, 7, 7	0	7, 8, 8, 8
2, 3, 3, 4, 8, 9, 9, 9	1	2, 3, 3, 5, 5, 5, 7, 7, 8, 9
3, 4, 5, 6, 7, 8, 9	2	0, 0, 2, 3, 5, 9
2, 3, 8, 9	3	1, 1, 9
0	4	

(b)

	Lowest	Q1	Median	Q3	Highest
7X	4	13	23	298	40
7Y	7	13	17	23	39

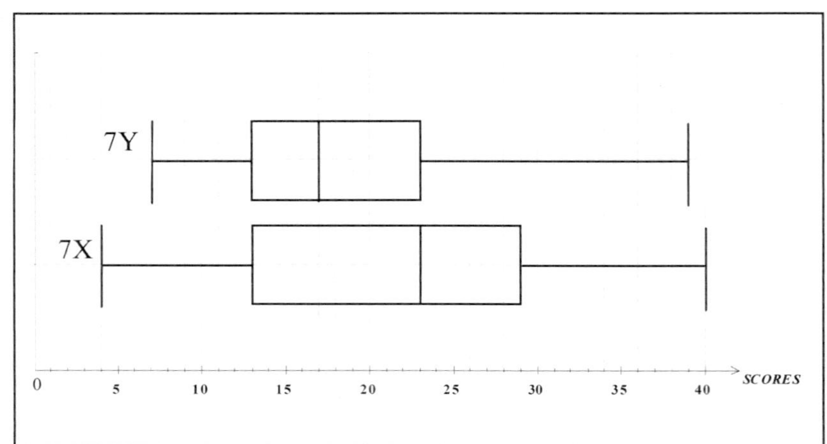

(c) The results from 7X are more spread out but generally higher that 7Y.

2. (a)

Interval	Tally	Frequency
0 – < 2	|	1
2 – < 4	|||| |	6
4 – < 6	||||	5
6 – < 8	|||	3
8 – < 10	|||| |	6
10 – < 12	|||	3
12 – < 14	||||	4
14 – 16	||	2

(b)

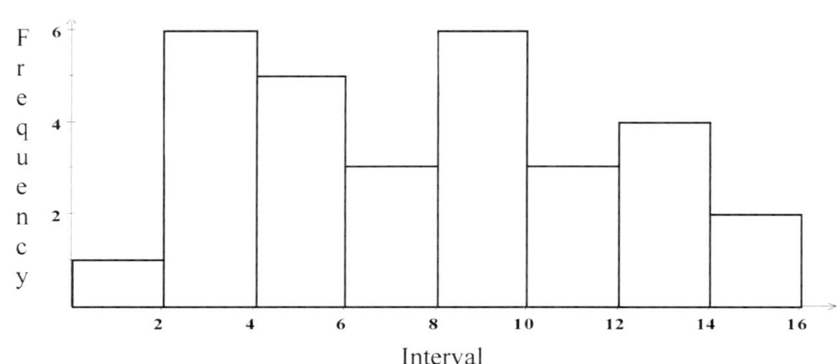

(c) Angles in table order: 12°, 72°, 60°, 36°, 72°, 36°, 48°, 24°

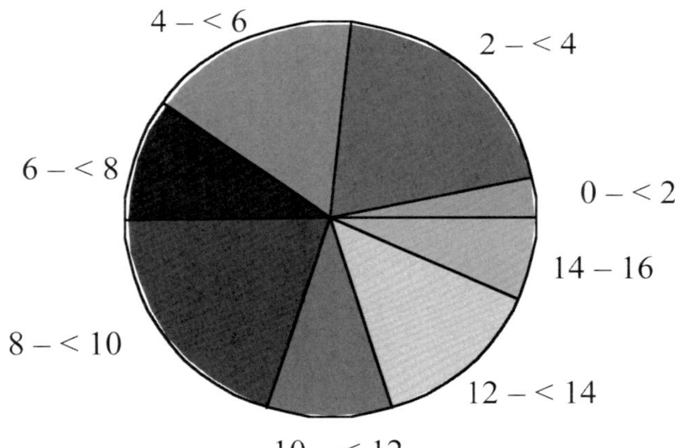

(d) 20

(e) 9

(f) $\frac{13}{30} \times 100 = 43\frac{1}{3}\%$

(g) $100 - 43\frac{1}{3} = 56\frac{2}{3}\%$

(h) Only 30% of the plants in the trial failed to reach 5.3cm under controlled lighting conditions. Based on this result it would appear that the plants in the trial grew better than the one grown outside.

The experiment needs to be expanded to draw more accurate conclusions.

Chapter 9

Probability

Areas of Interest

Games

Probability with Order

Using Diagrams

Unfair Dice

Venn Diagrams

Probability and Statistics

Games

1. Lleyton's chance of winning a game is $\frac{2}{3}$ when he plays Roger.

   What is the probability that Lleyton wins a set:

   (a) 6 games to love (zero)
   (b) 6 games to 1

   (c) 6 games to 2
   (d) 6 games to 3?

2. A game with two players is played using two coins and a die.

   Player A starts by tossing both coins. If the coins show two heads or two tails that player continues to play, otherwise the other player takes over. The die is rolled by the active player. If the number is three or less, the player gets a point but if the number is 4 or more, the player loses a point. The same player tosses the coins and the game continues. The game continues until the difference between the scores is three and therefore the player with the greater score wins.

   (a) Describe what has happened in the game and find the probability that games with the following scorecards will happen.

   (i)

   Game 1

Round	Player A	Player B
1	1	
2	0	
3		−1
4		−2
5	1	

   (ii)

   Game 2

Round	Player A	Player B
1		−1
2	1	
3		0
4	2	
5		−1

   (b) Play the game with a friend and comment as to whether the player to start the game has an advantage.

Probability with Order

1. Three women at a hospital have babies during a week. The chance of a male birth is $\frac{2}{3}$.
   Find the probability that:
   (a) only girls will be born

   (b) all children born are of the same sex

   (c) exactly two women will give birth to girls

   (d) at least one boy is born.

2. There are four coins in a bag, two double headers and two normal coins (with a head and a tail).
   (a) A coin is chosen at random from the bag and tossed. Find the probability that the coin shows:
   (i) a head (ii) a tail.

   (b) A coin is chosen and tossed and the result noted. It is put back into the bag and another coin chosen then tossed and the result noted. Find the probability that both results were:
   (i) heads (ii) tails (iii) a tail and a head.

Probability with Order page 2

2. (c) If two coins were chosen from the bag and tossed, find the probability that both coins showed:
   (i) heads (ii) tails (iii) a head and a tail.

3. A boy is presented with red, blue and white buzzers. He is given the task of pressing any of the buzzers three times in the correct order to receive a treat. Find the probability that:
   (a) he presses them in the correct order the first time

   (b) he does not press them in the correct order the first time.

   His friend is presented with four different coloured buzzers. She is given the same task of pressing any of the buzzers three times in the correct order to avoid extra homework. Find the probability that:
   (c) she presses them in the correct order the first time

   (d) she does not press them in the correct order the first time.

Using Diagrams

1. Two fair dice are rolled (red and blue).

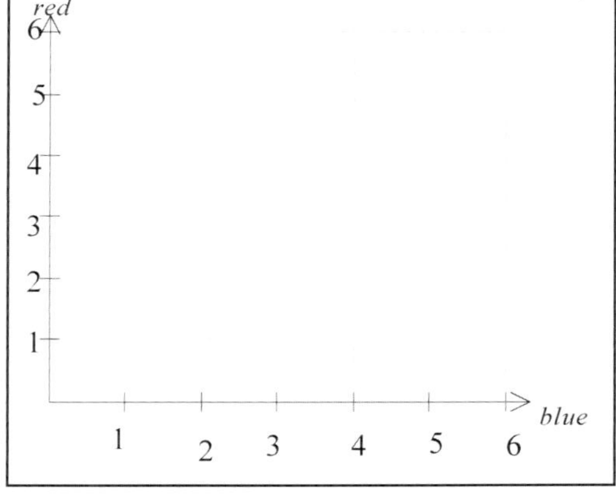

(a) (i) If the values on each die are to be multiplied, show appropriate values on the grid.

(ii) Find the probability that the product of the numbers shown on the dice is less than 12.

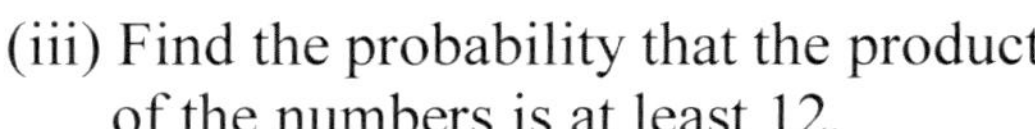

(iii) Find the probability that the product of the numbers is at least 12.

(iv) Find the probability that the product of the numbers is less than 20.

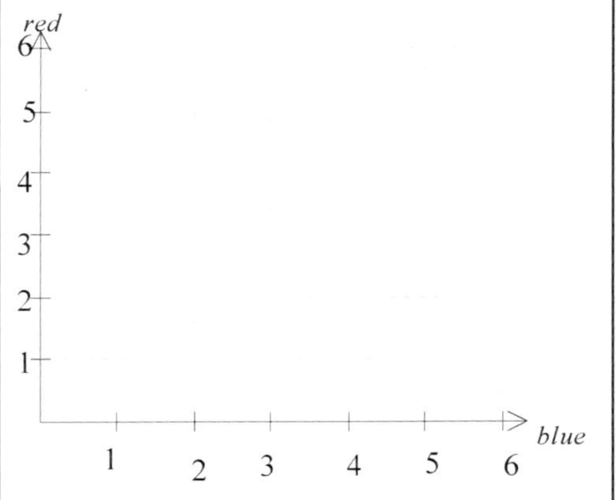

(b) (i) If the values on the grid are the difference between the numbers on the dice, show appropriate values on the grid.

Find the probability that the difference between the numbers is:

(ii) 2

(iii) 3

(iv) 0

(v) at least 4

(vi) at most 4.

Using Diagrams page 2

2. Three darts are thrown from a long distance to hit this dartboard at random.
   (a) Use a tree diagram to show the possibilities along with the probabilities of when the darts are thrown at the dartboard.

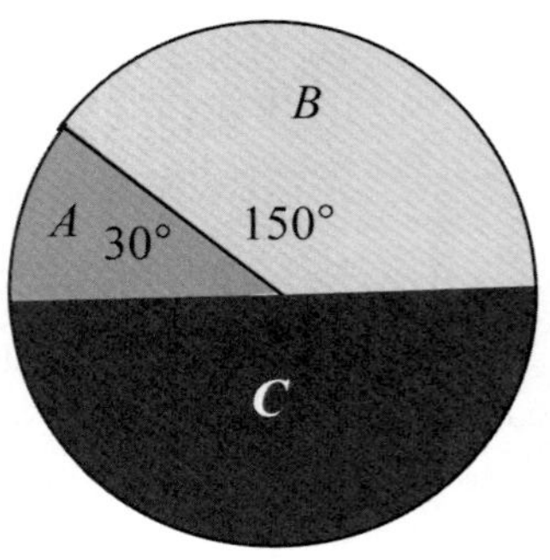

   (b) Find the probability that the darts will hit:
      (i) all three areas

      (ii) area B twice and area A once

      (iii) area B three times, given that area B is hit at least twice.

Unfair Dice

1. A gambler has two dice: a red four-sided die numbered (1, 2, 3, 4), where the $Pr(1) = 4\,Pr(4)$, $Pr(1)=2Pr(3)$, $Pr(2)=3Pr(4)$ and a fair blue five-sided die numbered (1, 2, 3, 4, 5).

   (a) Show the outcomes on a lattice diagram.

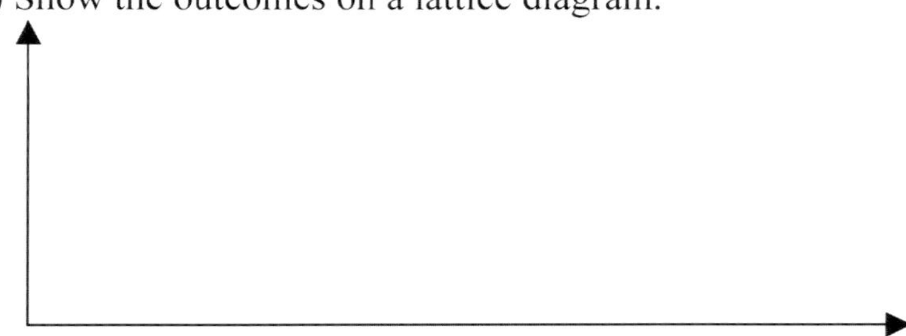

   The dice are rolled and the sum noted.

   (b) Find the probability that the sum of the two dice is larger than 6.

   (c) The gambler produces two more red and blue dice with the same numbers printed on each face as before. Each die is biased so that the probability of each face showing is proportional to the number on that face.

   (i) Complete the probability table for each die.

	1	2	3	4	5
red					
blue					

   (ii) Find the probability that the sum of the dice is less than 6.

2. A spinner at the local casino with the numbers 1, 2 and 3 on it is known to be loaded (unfair). It has been found that that the probabilities of spinning the numbers are as follows:

   $Pr(1) = Pr(2) = 0.4$ and $Pr(3) = 0.2$

   (a) The spinner is spun twice and the numbers noted. List the events and find the probability for each different sum.

   (b) If the sum of the dice is at least four, find the probability that the dice showed different numbers.

   (c) The spinner is spun twice. If the probability of the event is 0.16, what could the events have been?

Venn Diagrams

1. Two sets of numbers are shown on the Venn diagram below. State the elements in the following.

(a) $A \cup B$

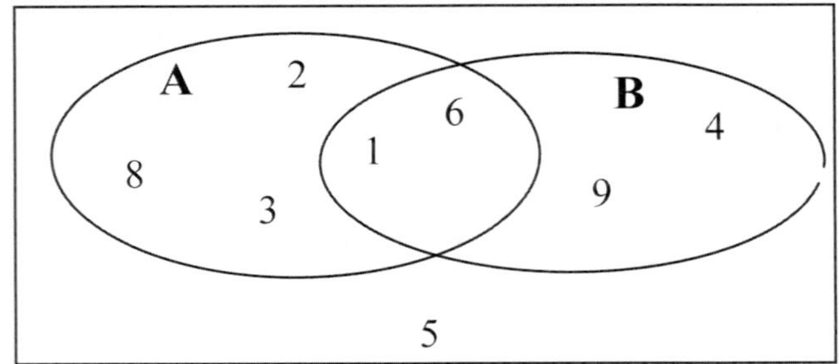

(b) $A \cap B$

(c) A'

(d) B'

(e) $(A \cap B)'$

(f) $(A \cup B)'$

2. Two sets of numbers are: A: {0, 1, 5, 7, 8, 12, 16} B: {2, 4, 6, 8, 10, 12, 14, 16} where $\xi = \{0,1,2,3,4,5,6,7,8,9,10,11,12,13,14,15,16\}$

(a) Complete the information on the Venn diagram.

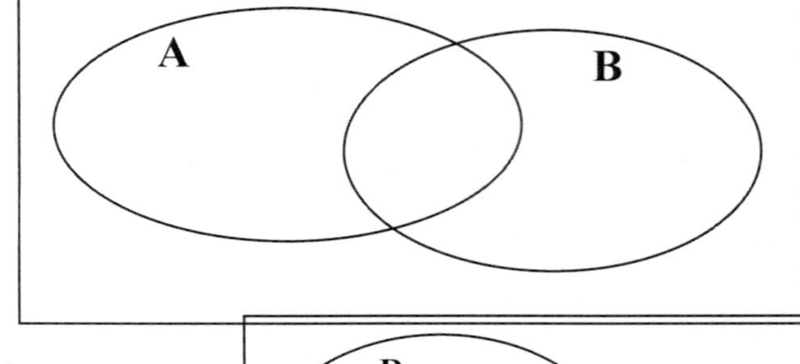

(b) Use the diagram, or otherwise, to find:

(i) $A \cap B$
(ii) $A \cup B$
(iii) $A' \cap B$
(iv) $A \cup B'$

3. Students were surveyed as to which fast food they enjoyed from burgers, chips and pizzas.
11 students liked chips. 11 students liked pizza whilst 5 students liked pizza only.
10 students liked burgers and chips and 3 liked all three types of fast food. 24 students liked burgers and 5 liked burgers and pizza.
4 students liked chips and pizza whilst 3 did not like any of the choices.

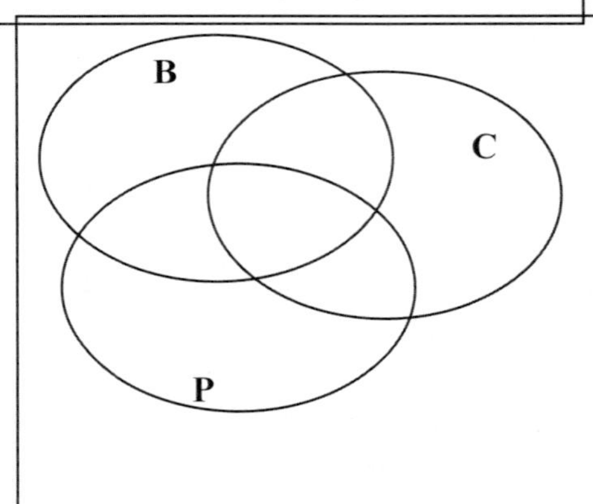

(a) Complete the Venn diagram.

(b) If a student was chosen at random find the probability that they:

(i) liked burgers but not pizzas
(ii) like chips but not pizza.

(c) Given that the person likes burgers, what is the probability that they do not like chips.

Probability and Statistics

1. Two equally biased coins are tossed one hundred times with the following results.

Number of heads	0	1	2
Frequency	64	32	4

(a) Find the probability of tossing:
(i) a head (ii) a tail with a single coin.

(b) If both coins were tossed twice, find the probability of obtaining:
(i) all tails (ii) one head and three tails.

(c) A coin is tossed three times. In one trial, a tail is observed to happen at least twice. Find the probability that three tails occurred.

2. A spinner is divided into three sectors on a circular spinner with areas labelled *A, B*, and *C*. When it is spun 250 times, the following results occur.

Area on spinner	*A*	*B*	*C*
Frequency	32	156	62

(a) If the spinner is divided into eighths, find the long-term probability for each area.

(b) If the spinner is spun six times, find the probability that it lands:
(i) twice on each area

(ii) once on *A*, twice on *B* and three times on *C*.

(c) The spinner is altered so that when it starts from *C* it is impossible to finish at *C* and the chances of the spinner finishing on *A* and *B* is in the ratio 3:1.
(i) Describe the way that the probabilities for *A* and *B* have changed from the original spinner.

The spinner is picked up at random and spun. Find the probability that it lands on:
(ii) *A* (iii) *B* (iv) *C*

Solutions

Games

1. (a) $\frac{64}{729}$ (b) $\frac{128}{729}$ (c) $\frac{448}{2187}$ (d) $\frac{3584}{19683}$
2. (a) (i)
   Player A: Coins – (T,T) or (H,H) to continue playing, then Die – {1, 2 or 3} to take 1 point
   Player A: Coins – (T,T) or (H,H) then Die – {4, 5 or 6} to lose 1 point
   Player A: Coins – (T,H) or (H,T) Player B to play. Die – {4, 5 or 6} to lose 1 point
   Player B: Coins – (T,T) or (H,H) then Die – {4, 5 or 6} to lose 1 point
   Player B: Coins – (T,H) or (H,T) Player A to play. Die – {1, 2 or 3} to take 1 point
   Difference is 3 points with Player A being in front: Player A wins.
   (ii)
   Player A: Coins – (T,H) or (H,T) Player B to play. Die – {4, 5 or 6} to lose 1 point
   Player B: Coins – (T,H) or (H,T) Player A to play. Die – {1, 2 or 3} to take 1 point
   Player A: Coins – (T,H) or (H,T) Player B to play. Die – {1, 2 or 3} to take 1 point
   Player B: Coins – (T,H) or (H,T) Player A to play. Die – {1, 2 or 3} to take 1 point
   Player A: Coins – (T,H) or (H,T) Player B to play. Die – {4, 5 or 6} to take 1 point

Probability with Order

1. (a) $\frac{1}{27}$ (b) $\frac{1}{3}$ (c) $\frac{2}{9}$ (d) $\frac{26}{27}$
2. (a) (i) $\frac{3}{4}$ (ii) $\frac{1}{4}$
   (b) (i) $\frac{9}{16}$ (ii) $\frac{1}{16}$ (iii) $\frac{3}{8}$
   (c) (i) $\frac{9}{16}$ (ii) $\frac{1}{16}$ (iii) $\frac{3}{8}$
3. (a) $\frac{1}{27}$ (b) $\frac{26}{27}$ (c) $\frac{1}{64}$ (d) $\frac{63}{64}$

Using Diagrams

1. (a) (i)

red \ blue	1	2	3	4	5	6
6	6	12	18	24	30	36
5	5	10	15	20	25	30
4	4	8	12	16	20	24
3	3	6	9	12	15	18
2	2	4	6	8	10	12
1	1	2	3	4	5	6

   (ii) $\frac{19}{36}$ (iii) $\frac{17}{36}$ (iv) $\frac{28}{36} = \frac{7}{9}$

1. (b) (i)

red \ blue	1	2	3	4	5	6
6	5	4	3	2	1	0
5	4	3	2	1	0	1
4	3	2	1	0	1	2
3	2	1	0	1	2	3
2	1	0	1	2	3	4
1	0	1	2	3	4	5

(ii) $\frac{8}{36}=\frac{2}{9}$ (iii) $\frac{6}{36}=\frac{1}{6}$ (iv) $\frac{6}{36}=\frac{1}{6}$ (v) $\frac{6}{36}=\frac{1}{6}$ (vi) $\frac{34}{36}=\frac{17}{18}$

2. (a)

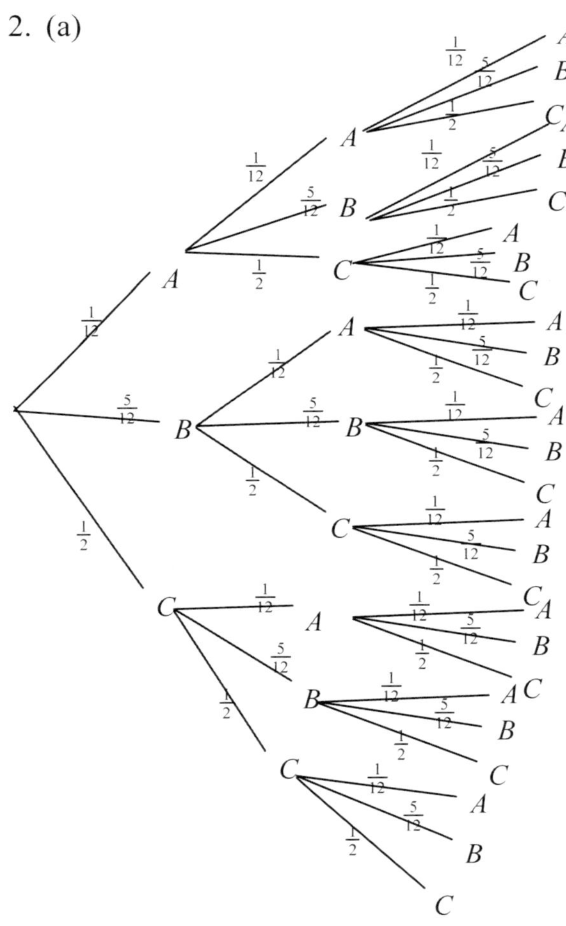

(b) (i) $\frac{5}{48}$ (ii) $\frac{25}{576}$

(c) $\frac{5}{21}$

Unfair Dice

1. (a)

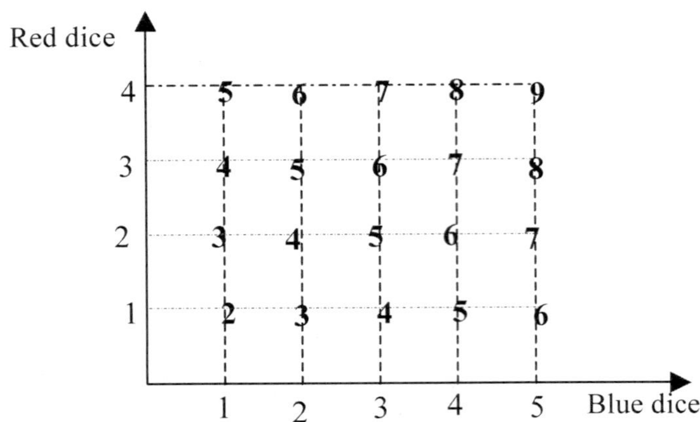

(b) $\frac{1}{5}$

(c) (i)

	1	2	3	4	5
red	$\frac{1}{15}$	$\frac{2}{15}$	$\frac{3}{10}$	$\frac{2}{5}$	
blue	$\frac{1}{15}$	$\frac{2}{15}$	$\frac{1}{5}$	$\frac{4}{15}$	$\frac{1}{3}$

(ii) $\frac{7}{30}$

2. (a)

Sum	2	3	4	5	6
Events	(1,1)	(1,2) (2,1)	(2,2) (1,3) (3,1)	(2,3) (3,2)	(3,3)
Probability	$0.4^2 = 0.16$	$2 \times 0.4 \times 0.4 = 0.32$	$0.4^2 + 2 \times 0.2 \times 0.4 = 0.32$	$2 \times 0.2 \times 0.4 = 0.16$	$0.2^2 = 0.04$

(b) $\frac{8}{13}$

(c) The events with a probability of 0.16 are: two 1's, two 2's, or a 1 and a 2 (in no particular order).

Venn Diagrams

1. (a) $\{1,2,3,4,6,8,9\}$ (b) $\{1,6\}$ (c) $\{4,5,9\}$ (d) $\{2,3,5,8\}$
(e) $\{2,3,4,5,8,9\}$ (f) $\{5\}$

2. (a)

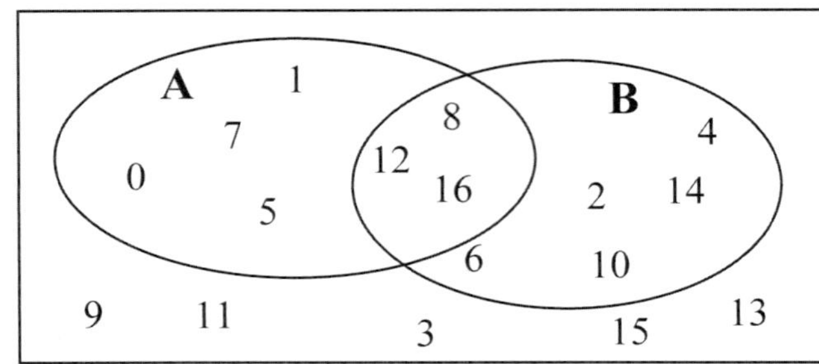

2. (b) (i) $\{8,12,16\}$ (ii) $\{0,1,2,4,5,6,7,8,10,12,14,16\}$ (iii) $\{2,4,6,10,14\}$
(iv) $\{0,1,3,5,7,8,9,11,12,13,15,16\}$

3. (a)

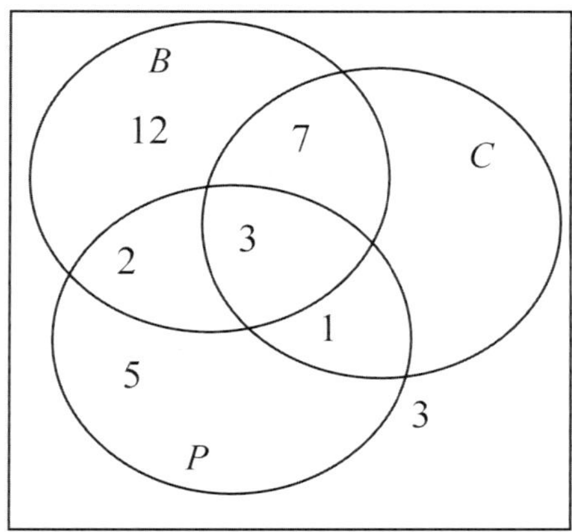

(b) (i) $\frac{12}{33}$ (ii) $\frac{7}{33}$ (c) $\frac{7}{12}$

Probability and Statistics

1. (a) (i) Pr(head) = $\frac{1}{5}$ (ii) Pr(tail) = $\frac{4}{5}$ (b) (i) $\frac{256}{625}$ (ii) $\frac{256}{625}$

   (c) $\frac{4}{7}$

2. (a) $\Pr(A) = \frac{1}{8}$, $\Pr(B) = \frac{5}{8}$, $\Pr(C) = \frac{1}{4}$

   (b) (i) $\frac{1125}{32768}$ (ii) $\frac{135}{8192}$

   (c) (i) The probability for landing on A and B has not changed when the spinner is spun from A or B.

   When the spinner starts from C, Event A has increased 6 times. Event B is $\frac{2}{5}$ times its original value.

   (ii) $\frac{9}{32}$ (iii) $\frac{17}{32}$ (iv) $\frac{3}{16}$

Chapter 10

Problem Solving

Areas of Interest

Logic (1) to (3)

Number (1) to (7)

Speed

Distance and Travel

Number and Algebra (1) & 2

Algebra

Shapes (1) & (2)

Space

Logic (1)

1. Jill, Jennifer and Janette are form teachers and they each teach a different subject. One of them teaches mathematics, and one is the form teacher for 10Y. The one in charge of Form 8Z teaches Science. Jennifer is in charge of Form 9X, and Janette teaches French. Find each teacher's subject and form.

2. How tall is a stick that is 4 metres shorter than a tree three times as tall as the stick?

3. Four flags are to be placed so that each flag is the same distance from the other three. Draw a diagram and describe how this can be done.

4. When Doug was 8, his father was 30. His father is now twice as old as Doug. Find the difference in their present ages.

5. Five years ago John was three times Mary's age. In ten years, he will be one-and-a-half times as old as Mary. How many times is John older than Mary now?

6. William walked 10km north, 10km east and them 10km south. He returned to where he started. Explain how this is possible.

Logic (2)

1. Complete the last figure in the sequence of shapes.

   (a)

   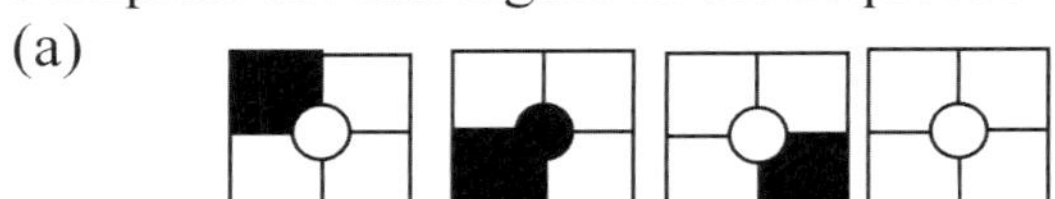

   (b)

   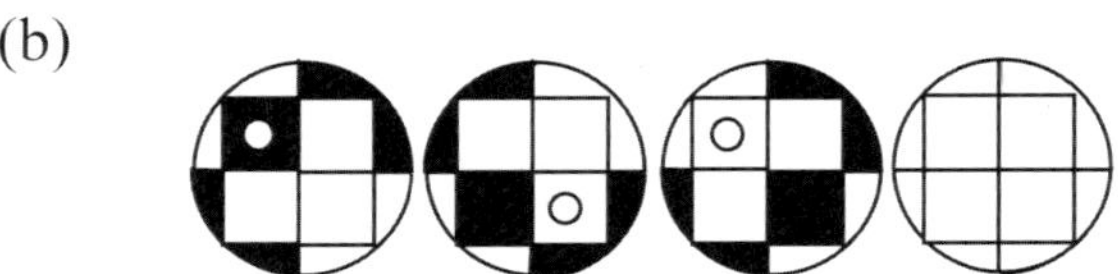

   (c)

   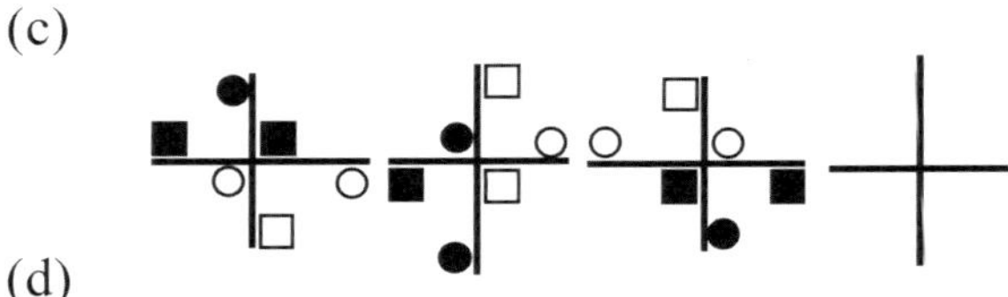

   (d)

2. This shape is made from eight matches pointing upwards. Move three matches to turn the shape upside down.

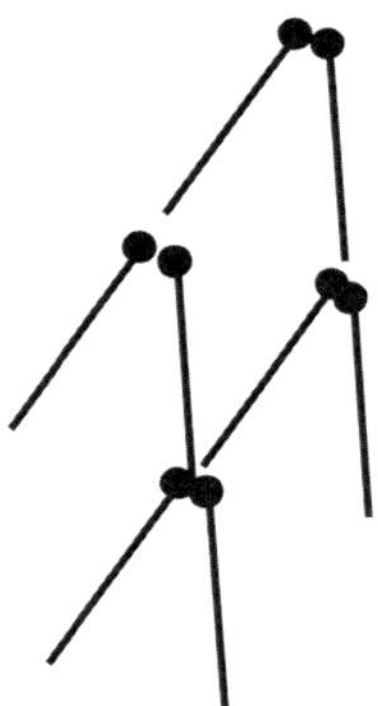

3. Draw four straight lines without lifting your pencil off the page to pass through the nine points shown in the diagram.

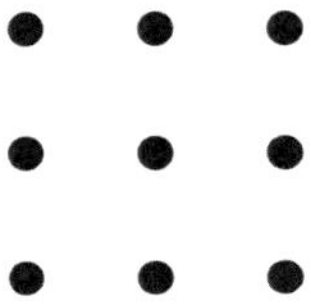

Logic (2) page 2

4. Weights A, B, X, Y and Z are shown in balance on four sets of scales (A to D). Which of the weights should be placed onto scale E to make it balance?

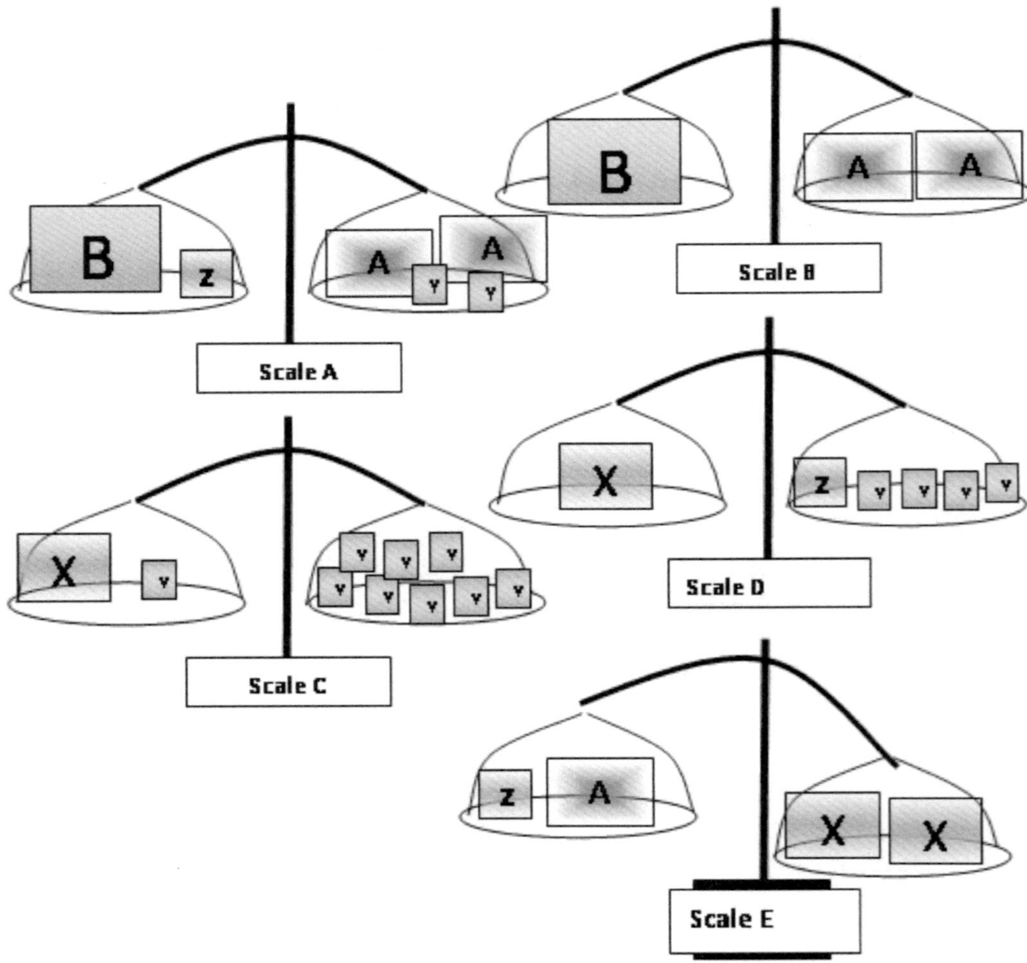

Logic (3)

1. During a split round in the AFL only four matches were played. The tips from three media experts were:
   Tipster A: Geelong, Essendon, Collingwood, Sydney
   Tipster B: Essendon, Geelong, Carlton, Melbourne
   Tipster C: Collingwood, Essendon, Carlton, St Kilda

   No one selected West Coast to win. Find the matches played during this split round.

2. Lucinda travels up 3 floors in a 15 floor, high-rise building, then down ten floors and finally up a number of floors to finish at the thirteenth floor. Describe the ways that she could have done this.

3. Alan, Bert, Cara and Doris each live in a house, which is brown, blue, white and blue but not necessarily in that order. They each have a pet different to the others and one travels to school by bus, another by train, another by bike and the last by walking.
   Use the following clues to match each person with the colour of their house, the pet they have and the way that they travel to school.

   The fish is kept in the white house and Alan keeps a dog in the black house. Cara has a rat for a pet and she walks to school. Bert does not live in a brown house but does ride his bike to school. Doris does not own a cat but travels to school by bus.

Logic (3) page 2

4. For the following questions, a weighing balance is to be used and all coins are of identical size.

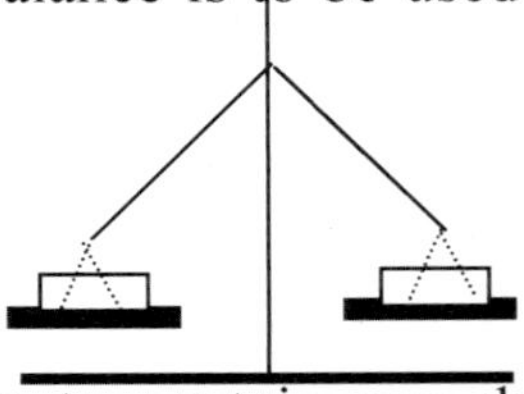

(a) A collection of three coins is known to contain one defective coin. If the defective coin is known to be underweight, then describe the most effective way that the coins could be weighed on a set of scales to locate the underweight coin.

(b) Extend the investigation to consider the situation of having one underweight coin in a collection of 4, 5, 6, 7, 8 and 9 coins. Find the possible ways that the coins could be weighed to find the underweight coin.

(c) A collection of three coins is known to contain one coin, which is either underweight or is overweight. Find the ways that the coins could be weighed to judge the weight of each coin.

(d) A collection of four coins is known to have a coin, which is either lighter or heavier than the other three coins. Find the ways that the coins could be weighed to determine the weight of each coin.

Number (1)

1. Doug is numbering the pages of a book in the way $\{1, 2, 3, \ldots\}$. Altogether he wrote 642 digits. How many pages are in the book?

2. In a 380-page book, how many times is the numeral 2 used?

3. A mathematics test consists of ten questions. Ten points are given for each correct answer and 3 points are deducted for an incorrect answer. If Ralph scored 61, how many did he get right?

Number (1) page 2

4. Place the numbers 1, 2, 4, 8, 16, 32, 64, 128 and 256 in a 3 x 3 square grid so that the numbers in each row, each column and each diagonal when multiplied together are equal to the cube of the number that is in the centre of the grid.

(a)

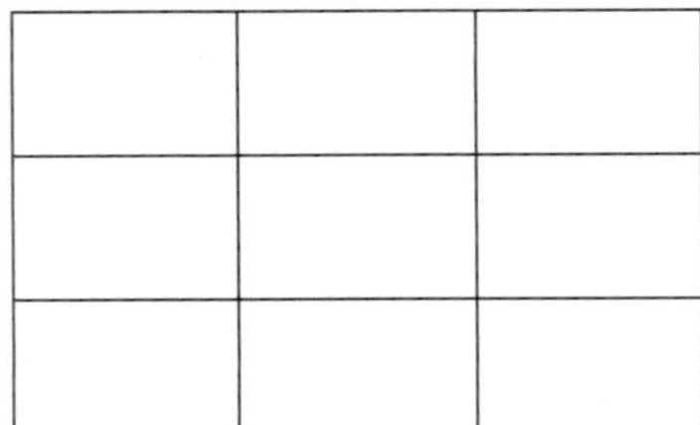

(b) Complete the grid converting the numbers into powers of 2.

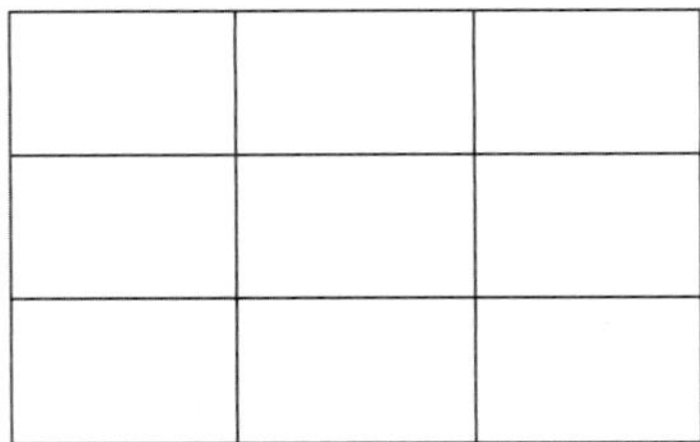

5. Write all the numbers from 1 to 12 using five twos (i.e. 2, 2, 2, 2, 2) and any mathematical operations you want.
(E.g. $10 = 2 + 2 + 2 + 2 + 2$)

6. A two-digit number is one more than eight times the sum of its digits. If the digits are reversed, the new number is four more than twice the sum of the digits. Find the numbers.

Number (2)

1. In a factory the average percentage of male employees absent each day over the 1990 year was 7%, and the mean percentage of female employees absent each day was 2%. If there are four times as many males as females in the factory, find the overall absentee percentage for the factory.

2. A Department of Conservation scientist caught 100 trout in various parts of a small lake. She tagged them and then threw them back in the lake. A week later 250 trout were caught and 35 of them had tags. Estimate the number of trout in the lake justifying how it is made according to the level of significance.

3. $\frac{1}{4}$ = ...% of $\frac{1}{3}$.

4. Dalia agrees to do Mirek's tax return if she can keep 10% of the tax refund obtained. In the end, Mirek received \$99. How much was the refund?

5. Ten stations lie at equal distances from each other along a railway line. How many times further is it from the first to the ninth than from the first to the third?

6. The gravitational attraction between two objects is given by the formula $F = \frac{Gm_1m_2}{d^2}$ where m_1 and m_2 represents the masses of the two objects, d is the distance between them and G is the gravitational constant where, $G = 6.67 x\ 10^{-11} m^3kg^{-1}s^{-2}$.

   (a) Calculate the gravitational force between two objects (m_1 = 480 000 kg, m_2 = 1 365 000 kg) that are separated by 200 metres.

   (b) What will be the effect on the force between the bodies if the mass of one body is doubled, the mass of the other body is tripled and the distance between them halved?

Number (3)

1. Lucinda was given \$8 for every correct problem she had on her homework. She lost \$5 for every question that she got wrong. After completing one homework sheet containing 26 questions she did not earn nor lose any money. How many questions did she get right?

2. Amy in her latest game of darts scored 199 points, which raised her average score from 177 to 178.
   (a) Find the number of dart games that she has now played.

   (b) What should she score in her next game to raise her average to 179?

3. In a yacht race the second leg was $\frac{2}{3}$ of the length of the first leg, the third leg was $\frac{3}{4}$ of the length of the second leg and the final leg was three times the length of the third leg. If the total distance of the race was 132km, how long was the final leg?

4. Bacteria double every minute. At 8am a single bacterium is placed in a large incubator. At 8:24a.m., the incubator is full. Find the time that the incubator is:
   (a) one quarter full

   (b) half full.

5. In Lucky Valley their numbers work in base 4 and they only use the digits {0, 1, 2, 3}. The first six numbers in ascending order are: {1, 2, 3, 10, 11, 12). Using this system their number 12 (in base 4) is the same as our number 6 (in base 10). Here is how it works: $12_4 = 1 \times 4^1 + 2 \times 4^0 = 4 + 2 = 6_{10}$. Find the Lucky Valley numbers for these numbers in base 10: 21, 37, 112, 272, 1 253 and 4 095.

Number (4)

1. The lighthouse at Point Lonsdale flashes every 9 minutes and the lighthouse at Point Nepean flashes every 12 minutes. If both flashed together at 6 p.m.:
   (a) list them until midnight.

   (b) how many times will they flash together until midnight?

2. In the Ventnor Channel (Phillip Island, Victoria) there are three buoys with flashing lights. The first buoy flashes every 3 minutes, the second flashes every 4 minutes and the third buoy flashes every 9 minutes. If both flashed together at 6 p.m., list the times that they will all flash together until midnight.

3. What fraction when it is expressed as a repeating decimal starts:
   (a) 0.204081632...

   (b) 0.103092781...?

4. Find as many pairs of positive whole numbers as you can that fit this equation:

$$\boxed{2p + 3q = 35}$$

Number (4) page 2

5. For

$13^2 = 169 \quad 133^2 = 17\ 689 \quad 1\ 333^2 = 1\ 776\ 889$

$16^2 = 256 \quad 166^2 = 27\ 556 \quad 1\ 666^2 = 2\ 775\ 556$

(a) without using a calculator, write down the answer to:

(i) $13\ 333^2$

(ii) $133\ 333^2$

(iii) $1\ 333\ 333^2$

(iv) $16\ 666^2$

(v) $166\ 666^2$

(vi) $1\ 666\ 666^2$

(b) Find another number between 10 and 20 that, if you keep repeating the second digit and squaring, has the same number pattern and write down some examples.

6. (a) Find the whole number that the series: $\frac{1}{2} + \frac{2}{4} + \frac{3}{8} + \frac{4}{16} + \frac{5}{32} + \ldots$ approaches by evaluating the sums of the first eight fractions.

(b) Find the whole number that the series: $\frac{1}{2} + \frac{3}{4} + \frac{5}{8} + \frac{7}{16} + \frac{9}{32} + \ldots$ approaches using a spreadsheet.

Number (5)

1. Remove exactly six digits so that the sum of the remaining numbers is:

(a) 356 (b) 146 (c) 3 336

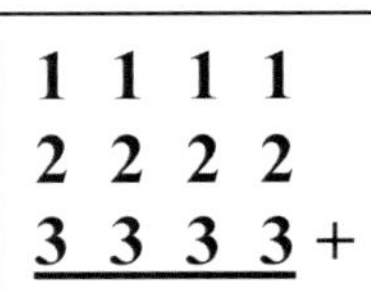

(d) 333 (e) 3 355 (f) 1 133

(g) 3 336 (h) 136 (i) 346

(j) 1 144

2. Find the digits, which can be used to form numbers, which result in the following additions.
Find 15 different answers for (b).

(a)

```
  T W O
  T O O
    T O +
  -----
  B E E
```

(b)

```
  O N E
  O F F
  T E N +
  -------
  N I N E
```

Number (5) page 2

3. Place the numbers in each circle so that no line joins consecutive numbers.

(a) Use numbers 1 to 5.

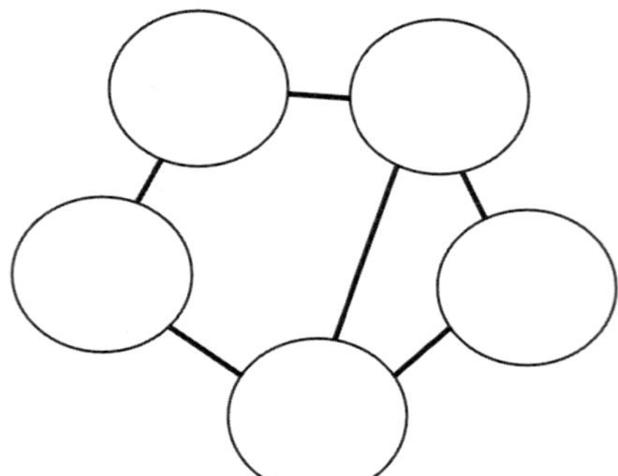

(b) Use numbers 1 to 6.

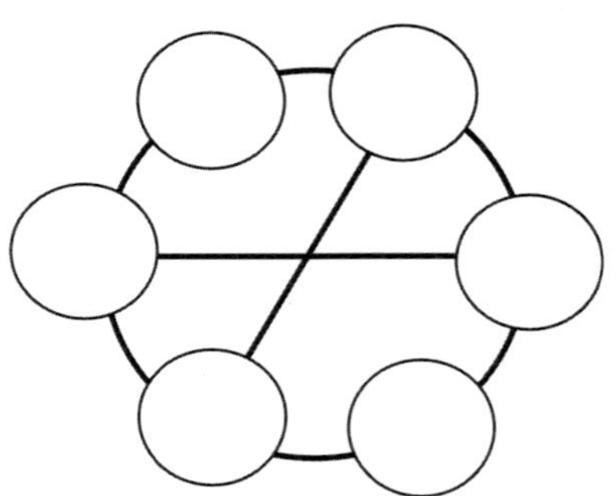

(c) Use numbers 1 to 7.

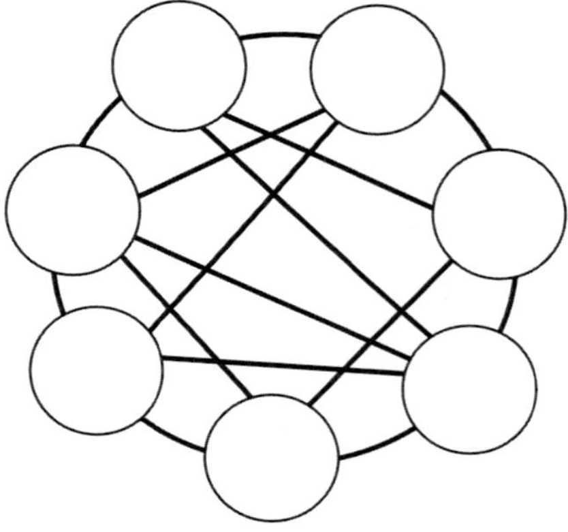

Number (6)

1. In a family, each boy has as many brothers as he has sisters. In the same family, each sister has twice as many brothers as she has sisters. Find the number of boys and girls in the family.

2. A group of seagulls were feeding at the local municipal tip when earthmoving equipment disturbed them. Initially, half of the seagulls were scared away and then 10 returned. This happened five more times – half the number of seagulls flying away and 8, then 6, 4 and finally 2 returning. If at the end of this process 8 seagulls remain, find the number of seagulls that were initially feeding at the tip.

3. When trying to guess Sam's house number Will said 207, Matthew guessed 219, Isabelle said 317 and Jenny 205. Sam told them that each person had exactly one digit correct in the correct position. What is Sam's house number?

4. Whilst visiting the Post Office, I purchased $2 worth of stamps. I bought some five-cent stamps and ten times as many two-cent stamps as five-cent stamps. I also bought some 10-cent stamps. Find the number of each value of stamp that I bought – there are three ways that it can be done.

5. (a) Find the original smallest whole number and the number which remains so that when:
   (i) one half is removed, then a third and finally a quarter is removed in that rder a whole number remains

   (ii) one half is removed, then a quarter and finally a third is removed in that order a whole number remains

   (iii) one third is removed, then a quarter and finally a third is removed in that order a whole number remains.

   (b) Find the smallest whole number greater that 100, so that after a third is removed, then a quarter and finally a half of a whole number remains. State an attribute of the remaining number.

Number (7)

1. Initially two numbers are in the ratio 5:1. After 6 is added to both numbers the ratio of the numbers is a:1.
Find the original numbers for the following values of a, and verify each solution.

(a) 2

(b) 3

(c) 4

2. When an aluminium extension ladder has five rungs showing, it stands 2.4 metres tall.
(a) Find the height of the ladder when the following numbers of rungs are showing.
(i) 8 (ii) 10

(iii) 12 (iv) x

(b) Find the total length of aluminium tubing required to make each of the ladders in part (a) if the width of each rung is 30cm.

3. Without evaluating the following, state with reasons which expressions have the same value.
2^{60} 4^{24} 8^{20} 16^{12} 32^{12} 64^{8}

4. Aaron and Beatrice start running from the same point, at constant speed, around a circular sporting field in opposite directions. They meet for the first time after Aaron has run 600 metres. When they meet for the second time, Beatrice has 200 metres to go to complete her first circuit of the track.
Find:
(a) the total length of the track

(b) the ratio of the speed of the faster runner to the slower runner.

Speed

1. A truck is travelling at 60 km/h and a car is travelling at 90 km/h towards the truck. They are 10km apart. How far apart will they be one minute before they meet?

2. In a 21km marathon, Andrew ran for 2 hours and walked for one hour. If Andrew runs at 3 times the speed that he walks, what are his running and walking speeds?

3. A river is flowing at a speed of 5 km/h. A man caught a ferry which travelled upstream at a speed of 10 km/h. After the ferry had travelled 5km, the man realised he had lost his hat overboard, just after boarding the ferry. How long would it take the ferry to return so that the man could get his hat?

4. Two cars start driving towards each other from towns, which are 480km apart, at the same time. On average one car travels 26 km/h faster than the other. If they meet 3 hours and 20 minutes after starting, find the speed of each car.

Distance and Travel

1. Use this distance table to answer the following questions.

Interstate road distance from Melbourne (km)

To Sydney		*To Adelaide*	
Melbourne	0	Melbourne	0
Seymour	96	Melton	38
Euroa	150	Bacchus Marsh	53
Violet Town	166	Ballarat	114
Benella	193	Beaufort	161
Wangaratta	232	Ararat	204
Albury	305	Stawell	235
Holbrook	373	Horsham	301
Kyamba	419	Dimboola	338
Gundagai	493	Nhill	377
Yass	591	Kaniva	417
Gunning	632	Bordertown	461
Goulburn	680	Keith	506
Marulan	709	Coonalpyn	572
Mittagong	766	Tailem Bend	624
Sydney	875	Adelaide	735

(a) What are the distances between:

(i) Melbourne and Sydney

(ii) Melbourne and Adelaide

(iii) Melbourne and Gundagai

(iv) Melbourne and Kaniva

(v) Yass and Marulan

(vi) Stawell and Adelaide

(vii) Sydney and Holbrook

(viii) Keith and Melton?

(b) What is the total distance for the following journey?
Melbourne – Euroa – Seymour – Goulburn – Albury and back to Melbourne

(c) What is the total distance for the following journey?
Adelaide – Kaniva – Keith – Ballarat – Dimboola – Bordertown to finish in Melbourne

Distance and Travel page 2

1. (d) Complete these distance charts.

	Adelaide	Coonalpyn	Bordertown	Nhill	Stawell	Ballarat	Melbourne
Melbourne							*
Ballarat						*	
Stawell					*		
Nhill				*			
Bordertown			*				
Coonalpyn		*					
Adelaide	*						

	Sydney	Mittagong	Goulburn	Gundagai	Wangaratta	Euroa	Melbourne
Melbourne							*
Euroa						*	
Wangaratta					*		
Gundagai				*			
Goulburn			*				
Mittagong		*					
Sydney	*						

2. A group of four people want to cross a crocodile-infested river. All they have is a raft that can carry 100 kg at the most. Ian weighs exactly 85 kg. Jenny weighs 54 kg, William weighs 45 kg and Lucinda weighs 40 kg. Explain how they could all cross the river, using the fewest possible crossings (they can all row the boat).

Number and Algebra (1)

1. Place the digits 1, 2, 3, 4, 5, 6, 7 and 8 into the circles so that each digit is not connected to one consecutive to it (so that 2 cannot be connected to 1 or 3).

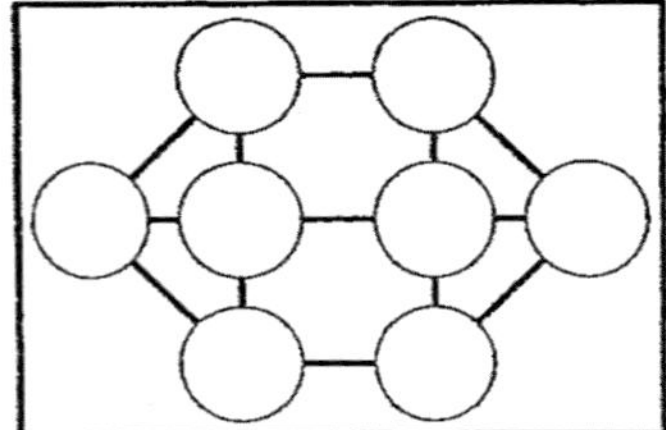

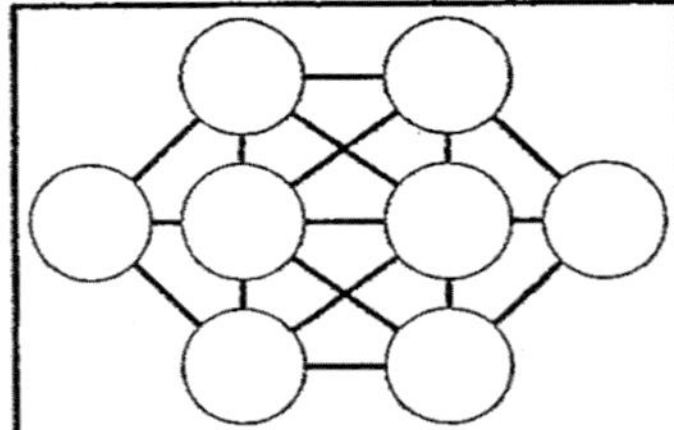

2. On a farm, there are a number of bulls and hens. The farmer counts a total of 14 heads and 48 legs. He also knows that all his animals are fit and well and still in possession of all their legs and heads. How many bulls and hens could there be on the farm?

3. In a jellybean guessing competition:
   Lucinda guessed 898 — Lucinda missed by 25
   Jenny guessed 890 — Jenny missed by 33
   William guessed 950 — William missed by 27
   How many jellybeans were in the jar? Explain how you got your answer.

4. Gordon likes to draw lines between points. Here are the drawings that he has made so far.

2 points	3 points	4 points	5 points	6 points
1 line	3 lines	6 lines		

   (a) Complete the lines drawn between 5 points and 6 points shown in the diagram. Count them as you go.

   (b) The number of lines drawn between these points makes a pattern. Identify the pattern and fill the table.

   (c) Use the number pattern to find the equation that connects the number of lines drawn between the points (L) and the number of points (P) in the pattern.

   (d) Use this relationship to predict the number of lines that will be drawn between:
   (i) 100 points (ii) 1 000 points.

No. of points	Lines
2	
3	
4	
5	
6	
7	
8	
9	
10	

Number and Algebra (2)

1. The sum of my two digits is even and the product of my digits is odd. If I am greater than 80, what numbers could I be?

2. I am a two-digit number. The sum of my digits is 11. If my digits were reversed, I would be 45 less than I am. What number am I?

3. (a) If 8 green crayons cost \$10, then how much will 13 of these crayons cost?

   (b) Red crayons cost \$1.30 and blue crayons cost \$1.45.

   (i) Find the cost of 20 red and 30 blue crayons.

   (ii) Express the cost (C) as an equation in terms of the number of red (r) and the number of blue (b) crayons.

   (iii) In a box of crayons, there are twice as many red ones as there are blue ones.If the box contains 1 500 crayons, find the number of red crayons and the number of blue crayons.

   (iv) Find the cost of this box of crayons.

Number and Algebra (2) page 2

4. Dots are arranged in a triangular pattern.

Row 1 (n = 1)	•	Pattern 1 (P = 1)
Row 1 (n = 1) Row 2 (n = 2)	• • •	Pattern 2 (P = 2)
Row 1 (n = 1) Row 2 (n = 2) Row 3 (n = 3)	• • • • • •	Pattern 3 (P = 3)

(a) How many dots are there in the:
(i) first row (ii) second row

(iii) third row?

(b) The row number is given the pronumeral n.
Complete the table.

(c) Write an equation that connects the number of dots in a row (d) to the row number (n).

Row No. (n)	No. of dots (d)
1	
2	
3	
4	
5	
6	
7	
8	
9	
10	

(d) The patterns are given names according to the number of rows that they have: $P = 1$, shows that the pattern has 1 row. Complete the table.

(e) Write an equation that connects the number of dots in the pattern (d) to the pattern number (P).

Pattern No. (P)	No. of rows	No. of dots (d)
1		
2		
3		
4		
5		
6		
7		
8		
9		
10		

(f) Use this equation to find the number of dots in a pattern with:
(i) 100 rows ($P = 100$)

(ii) 1 000 rows ($P = 1\ 000$)

Algebra

1. Use algebra to solve the following problems.
   (a) Twice the sum of three consecutive even numbers is 216. Find the numbers.

   (b) Three times the sum of two consecutive odd numbers is 264. Find the numbers.

2. For numbers x, $x+2$, $x+4$ where x is a positive whole number, use algebra to find the smallest set of numbers so that their sum is divisible by 5.

3. If a book's mass is 18 kilograms plus one-third of a book, what do three books weigh?

4. In Australian Rules Football, 6 points are given for a goal and 1 point is given for a behind. Find four sets of values for a and b so that the score of a goals b behinds has the same points value of ab.

5. If $12a+9b+20c=120$, $a+b+c=20$ and $b \in \{1,2,3,....45,50\}$, find six sets of whole number values of a,b and c.

Shapes (1)

1. This task is to find the total number of different-sized squares in the chessboard shown below.

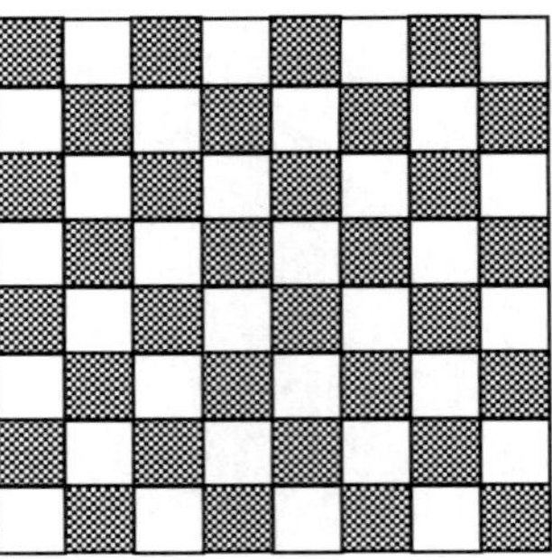

(a) Use the diagrams to find the total number of squares of all sizes in the following shapes.

(i) 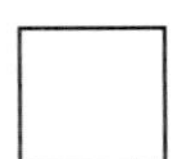

(ii) 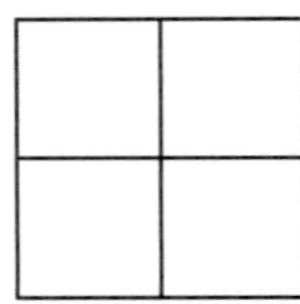

(iii) 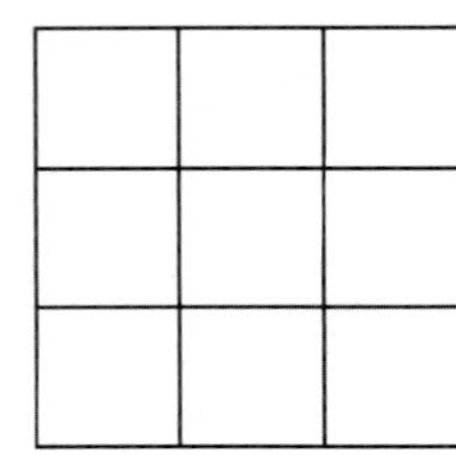

(b) Use the answers from part (a) to find the total number of squares in:

(i) 4×4 case

(ii) 5×5 case

2. (a) Complete your results in the table below continuing any patterns to that of 6×6, 7×7 and 8×8 cases.

Size of board	Number of squares
1×1	
2×2	
3×3	
4×4	
5×5	
6×6	
7×7	
8×8	

(b) Extend your investigation by finding an expression to find the total number of squares in an $n\times n$ board.

(c) Show by testing the formula for the cases: $n = 1, 2, 3, 4$ that the formula relating the total number of squares to the side length of the board is given by:

$$\textit{Number of squares} = \frac{n}{6}(n+1)(2n+1).$$

Shapes (2)

1. This task is to find the total number of ways that shapes can fit onto a chessboard without overlapping the lines. A standard 8×8 chessboard is shown below.

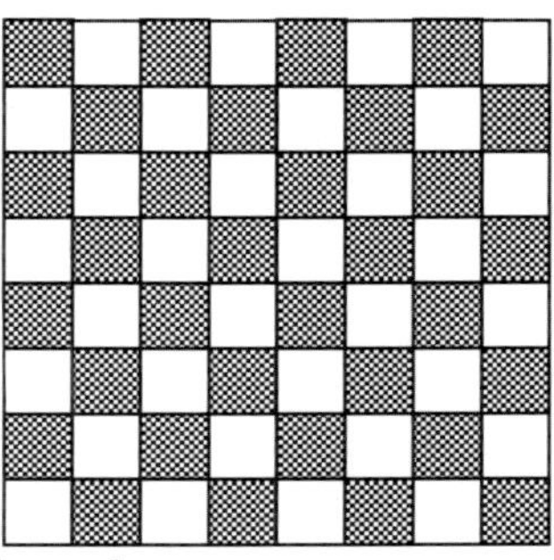

(a) (i) Explain why 64 single squares will fit onto the chessboard

(ii) Find the number of times that the 2×2 square shape will fit onto the board.

(iii) Find the number of times that the 3×3 square shape will fit onto the board.

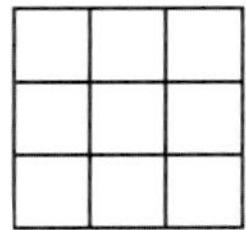

(b) Find the number of times that the following will fit onto the board.

(i) 4×4 shape (ii) 5×5 shape

2. Find the number of times that these shapes can be found on a chessboard.

(a)

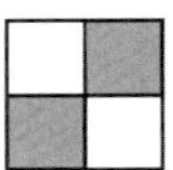

(b)

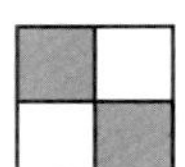

(c)

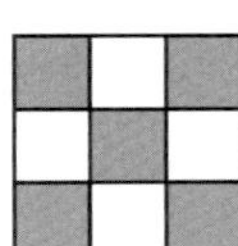

(d)

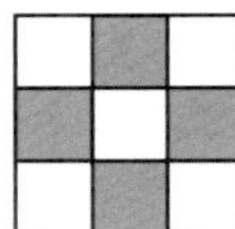

3. Find the number of times in terms of n that the following shapes can be placed on a chessboard size $n \times n$.

(a)

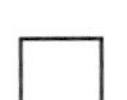

(b)

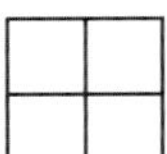

(c)

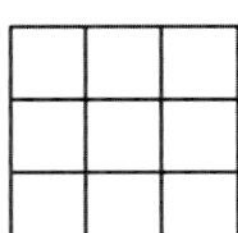

Shapes (2) page 2

4. Square pieces are to be placed onto the boards shown below.

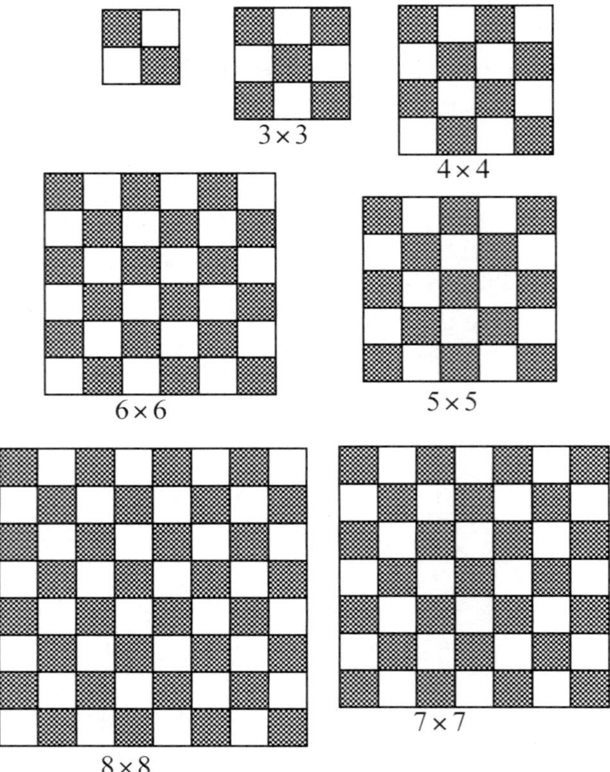

(a) (i) Find the number of times that this shape can be placed onto each board – complete the table.

Board size ($n \times n$)	Number of times shape fits on board N
2×2	0
3×3	$1 + 1 = 2$
4×4	$1 + 2 + 1 = 4$
5×5	
6×6	
7×7	
8×8	

Use the results in the table to show that the number of times N that the shape fits onto boards of size $n \times n$ is:

(ii) $N = \frac{1}{2}n^2 - n + \frac{1}{2}$, for n being odd

(iii) $N = \frac{1}{2}n^2 - n$, for n being even.

Shapes (2) page 3

4. (b) (i) Find the number of times that this shape can be placed onto each board– complete the table.

Board size ($n \times n$)	Number of times shape fits on board N
2×2	1
3×3	1 + 1 = 2
4×4	2 + 1 + 2 = 5
5×5	
6×6	
7×7	
8×8	

Use the results in the table to show that the number of times N that the shape fits onto boards of size $n\times n$ is:

(ii) $N = \frac{1}{2}n^2 - n + \frac{1}{2}$, for n being odd

(iii) $N = \frac{1}{2}n^2 - n + 1$, for n being even.

(c) (i) Find the number of times that this shape can be placed onto each board – complete the table.

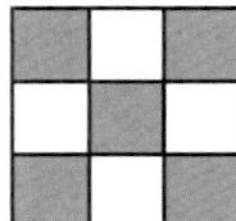

Board size ($n \times n$)	Number of times shape fits on board N
3×3	1
4×4	1 + 1 = 2
5×5	2 + 1 + 2 = 5
6×6	
7×7	
8×8	

Shapes (2) page 4

4. (c) Show that:

(ii) $N = \frac{1}{2}n^2 - 2n + 2\frac{1}{2}$, for n being odd

(iii) $N = \frac{1}{2}n^2 - 2n + 2$, for n being even.

(d) (i) Find the number of times that this shape can be placed onto each board – complete the table.

Board size ($n \times n$)	Number of times shape fits on board N
3×3	1
4×4	1 + 1 = 2
5×5	2 + 1 + 2 = 5
6×6	
7×7	
8×8	

Show that:

(ii) $N = \frac{1}{2}n^2 - 2n + 1\frac{1}{2}$, for n being odd

(iii) $N = \frac{1}{2}n^2 - 2n + 2$, for n being even.

Space

1. Four different coloured squares are to be positioned in a 2-by-2 grid so that there are two red squares, one blue square and one yellow square. The squares are to be arranged so that the red squares do not meet at any edge. Draw the four ways that they can be arranged.

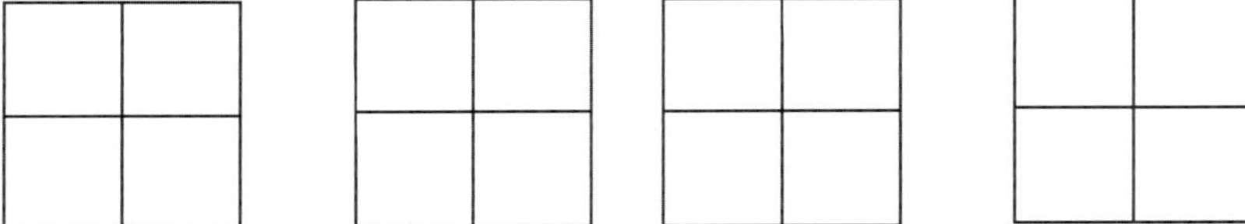

2. Draw the solid collection of cubes on the isometric grid using this plan and elevation information. The plan (base) is shown on the grid.

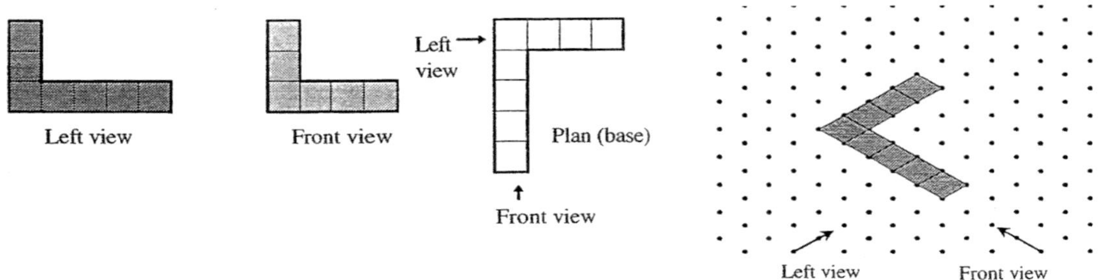

3. Blocks are placed on a 4 × 4 grid and viewed from positions A and B. You can use blocks to help you draw the diagrams.
   (a) Three blocks are placed onto the grid and when viewed from A and B they appear as shown. Draw the six possible positions of the blocks.

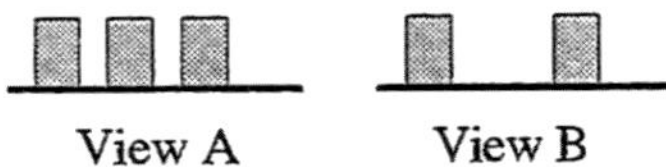

   (b) Four blocks are placed onto the grid and when viewed from A and B they appear as shown. Draw the 16 possible positions of the blocks.

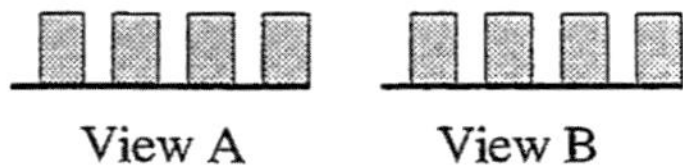

Measurement (1)

1. Three squares are shown with side lengths of 1cm, 2cm and 3cm respectively. Circles are drawn so that their perimeters are the same as each square.

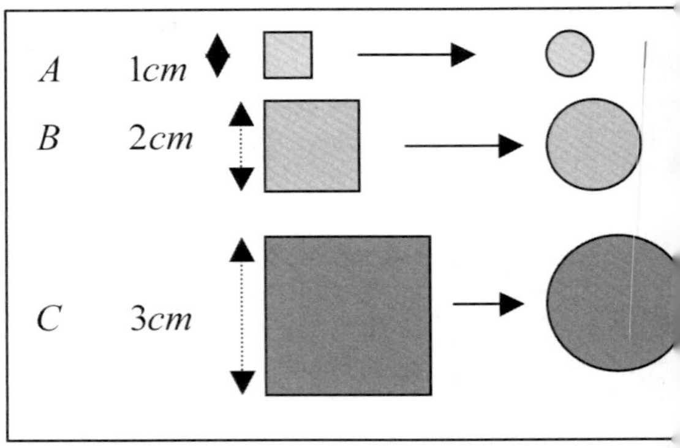

A	1*cm*	■	→	●
B	2*cm*	■	→	●
C	3*cm*	■	→	●

(a) Calculate the perimeters of each square.

(b) Find the radius of each circle.

(c) (i) Write an equation that expresses the radius of the circle (r) in terms of the edge length of the square (L) and π.

(ii) Use the equation to find the radius of a circle that has the same perimeter as a square of side length 10cm, expressing your answer to 2 decimal places.

(d) If another set of circles are drawn so that their areas are the same as the areas of the squares *A, B* and *C*.

(i) Write an equation that expresses the radius of the circle (r) in terms of the edge length of the square (L) and π.

(ii) Use the equation to find the radius of a circle that has the same area as a square of side length 2cm, expressing your answer in exact form and correct to 2 decimal places.

Measurement (1) page 2

2. Shown below is an equilateral triangle with side length 3cm.

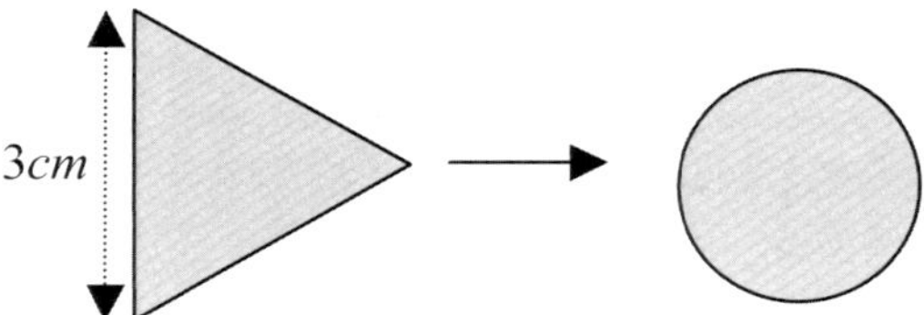

(a) Find the radius of the circle that is made from the triangle if their perimeters are the same.

(b) Write an equation that expresses the radius of the circle (r) in terms of the edge length of the equilateral triangle (L).

(c) Use the equation to find the radius of a circle that has the same perimeter as an equilateral triangle with side length 10cm, expressing your answer in exact form and correct to 2 decimal places.

Measurement (2)

1. Find the total number of squares of all sizes in the diagram.

2. Two ants start at A. One walks around a semicircle to B whilst the other walks around the smaller semicircles to finish at B. If in each case, the ants walk at the same speed which ant will reach B first?

(a)

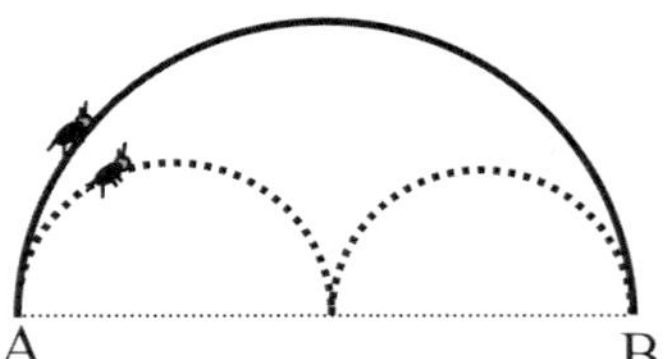

(b)

(c)

Measurement (2) page 2

3. Find the area of:
   (a) the rectangle in the quarter circle
   (b) the darker grey area in the quarter circle but not in the rectangle.

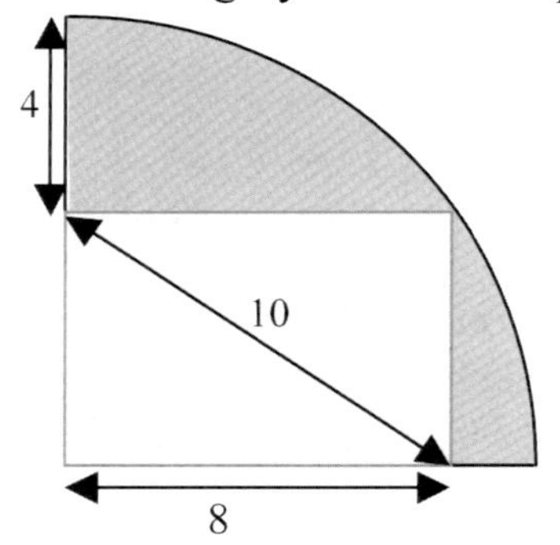

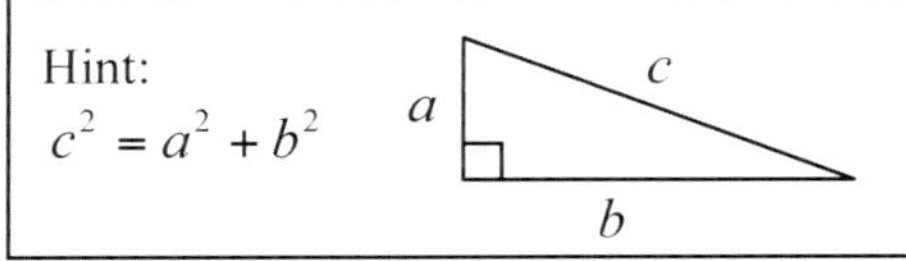

4. The following large triangular shape is made with four triangles and a rectangle.

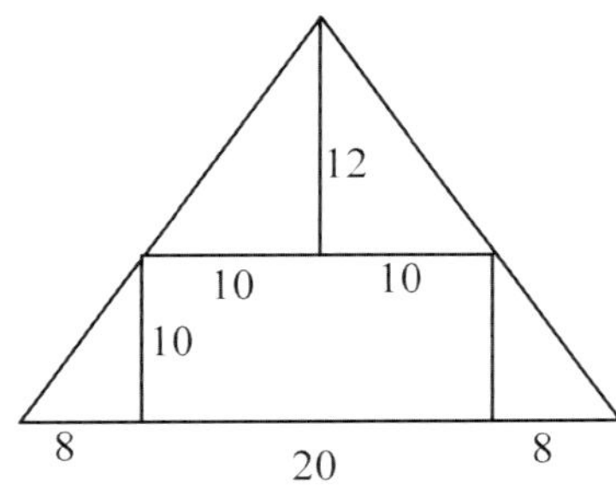

   (a) Find the area of the large triangle.

   (b) Find the area of each of the smaller shapes and add them up to find the area of the large triangle.

   (c) Explain why your answers to parts (a) and (b) are not the same.

   (d) Leaving the lengths of the rectangle unchanged; change a length to make the area of the large triangle the same as the sum of the individual.

Solutions

Logic (1)

1.

Jill	Jenny	Janette
Science	Maths	French
8Z	9X	10Y

2. 2 metres
3. tetrahedron
4. 22 years.
5. John's age is twice Mary's age
6. William started walking from the South Pole

Logic (2)

1. (a) 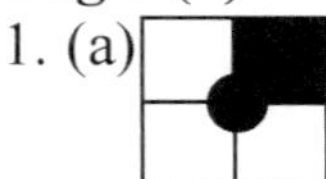 (b) 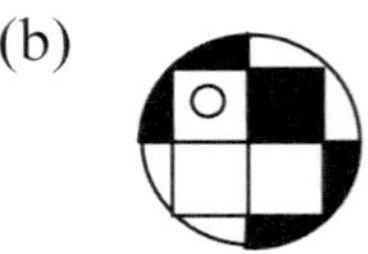

(c) 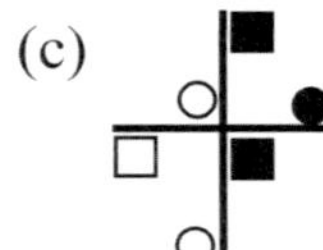 (d) 

2. 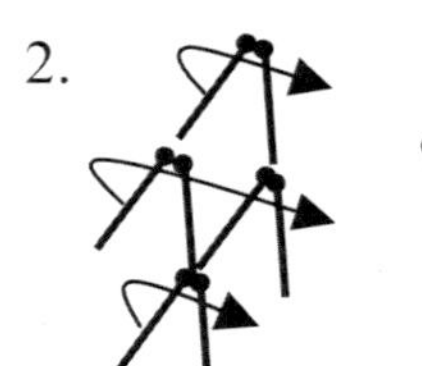 

3. 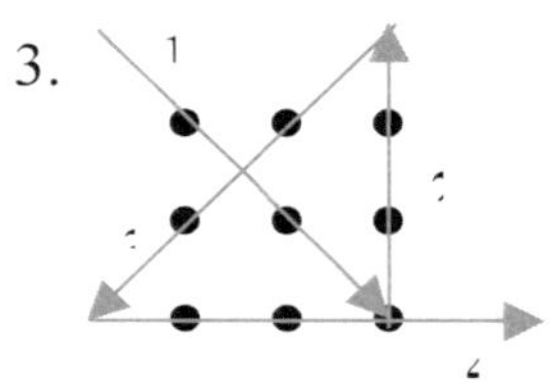

4.
One Y mass needs to placed on the left side of the scales.

Logic (3)

1. Essendon vs West Coast, Geelong vs St Kilda, Collingwood vs Melbourne, Sydney vs Carlton.

2.
Start on 7th floor go up 13 floors.
Start on 8th floor go up 12 floors.
Start on 9th floor go up 11 floors.
Start on 10th floor go up 10 floors.
Start on 11th floor go up 9 floors.
Start on 12th floor go up 8 floors.

3.

Alan	Dog	Black	Train
Bert	Cat	Blue	Bike
Cara	Rat	Brown	Walk
Doris	Fish	White	Bus

4. (a) 1 weighing
(b) 4 coins: 2, 5 coins: 1
6 coins: 2, 7 coins: 1, 8 coins: 3,
9 coins: 1 or 3
(c) 2 weighing
(d) Make a description

Number (1)

1. 250 pages

2. 86

3. 7 correct, 3 incorrect

4. (a) Product

32	64	2
1	16	256
128	4	8

(b) Indices (powers of 2)

5	6	1
0	4	8
7	2	3

5.

1	$2\times 2-2-\frac{2}{2}$
2	$2^2-2-2+2$
3	$\frac{2}{2}\times 2+\frac{2}{2}$
4	$\frac{2}{2}+\frac{2}{2}+2$
5	$\frac{2+2+2}{2}+2$
6	$2^2+2+2-2$
7	$2^2+\frac{2}{2}+2$
8	$2\times 2^2+2-2$
9	$2\times 2\times 2+\frac{2}{2}$
10	2+2+2+2+2
11	$\left(2+\frac{2}{2}\right)^2+2$
12	$(2^2)^2-2-2$
13	$\frac{(2^2)!}{2}+\frac{2}{2}$
14	$2\times 2\times 2^2-2$
15	$(2\times 2)^2-\frac{2}{2}$
16	$(2^2)^2+2-2$
17	$(2)^{2\times 2}+\frac{2}{2}$
18	$2\times 2^2\times 2+2$
19	$22-(2+\frac{2}{2})$ - this is cheating I can't find a solution to this question.
20	$(2^2)^2+2\times 2$
21	$(2\times 2)!-2-\frac{2}{2}$
22	$(2\times 2)!\times\frac{2}{2}-2$
23	$(2\times 2)!-2+\frac{2}{2}$
24	$\frac{1}{2}((2\times 2)!\times\sqrt{2\times 2})$
25	$(2^2)!+2-\frac{2}{2}$

6. 41 and 14

Number (2)

1. 6%
2. 710 to 720 (using 2 significant figures).
3. 75
4. $110
5. 4 times as far
6. (a) 0.001
   (b) The new force is 24 times that of the original force.

Number (3)

1. 10 correct questions
2. (a) played 22 games (b) score 201
3. 54km
4. (a) 22 mins (b) 23 mins
5. 111, 211, 1 300, 10 100, 103 211, 333 333

Number (4)

1. (a) They flash together every 36 minutes from 6p.m.: (6:36, 7:12, 7:48, 8:24, 9:00, 9:36, 10:12, 10:48, 11:24, 12 midnight).
   (b) ten times
2. They flash together every 30 minutes from 6p.m.: (6:00, 6:30, 7:00, 7:00, 7:30, 8:00, 8:30, 9:00, 9:30, 10:00, 10:30, 11:00, 11:30, 12 midnight).
3. (a) $\frac{6802721}{33333333}$ (b) $\frac{3436426}{33333333}$
4.

p	q
16	1
13	3
10	5
7	7
4	9
1	11

5.
   (a) (i) 177 768 889
   (ii) 17 777 688 889
   (iii) 1 777 776 888 889
   (iv) 277 755 556
   (v) 27 777 555 556
   (vi) 2 777 775 555 556
   (b) Powers of 19 behave in a similar way
6. (a) 2 (b) 3

Number (5)

1. (a)
```
    1
  2 2
3 3 3
-----
3 5 6
```
(b)
```
1 1 1
    2
  3 3
-----
1 4 6
```
(c)
```
      1
      2
3 3 3 3
-------
3 3 3 6
```
(d)
```
1 1 1
2 2 2
-----
3 3 3
```
(e)
```
    2 2
3 3 3 3
-------
3 3 5 5
```
(f)
```
1 1 1 1
    2 2
-------
1 1 3 3
```

1.(g)
```
   1
   2
3333
----
3336
```

(h)
```
111
 22
  3
---
136
```

(i)
```
 11
  2
333
---
346
```

(j)
```
1111
  33
----
1144
```

2. (a) $201 + 211 = 433$
   (b)
   $210 + 299 + 801 = 1310$,
   $310 + 399 + 501 = 1210$,
   $310 + 399 + 701 = 1410$,
   $310 + 399 + 801 = 1510$,
   $310 + 399 + 901 = 1610$,
   $410 + 499 + 301 = 1210$,
   $410 + 499 + 601 = 1510$,
   $410 + 499 + 701 = 1610$,
   $410 + 499 + 801 = 1710$,
   $410 + 499 + 901 = 1810$,
   $510 + 599 + 201 = 1310$,
   $510 + 599 + 301 = 1410$,
   $510 + 599 + 601 = 1710$,
   $510 + 599 + 701 = 1810$,
   $510 + 599 + 801 = 1910$,

3. (a)

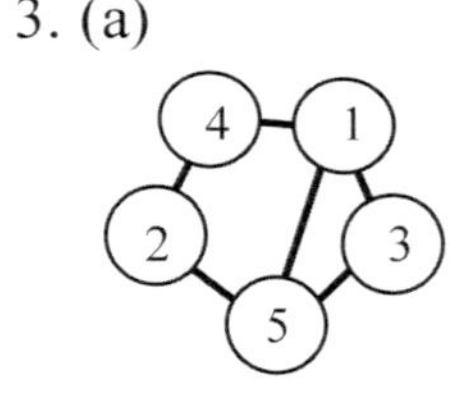

(b)

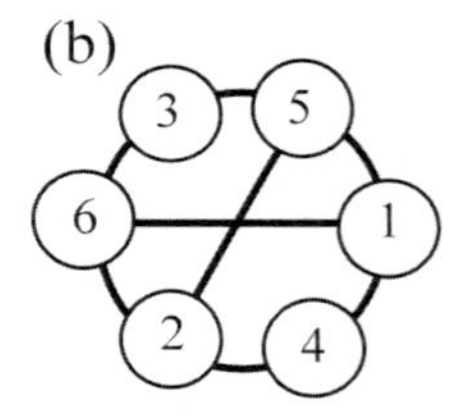

3. (c)

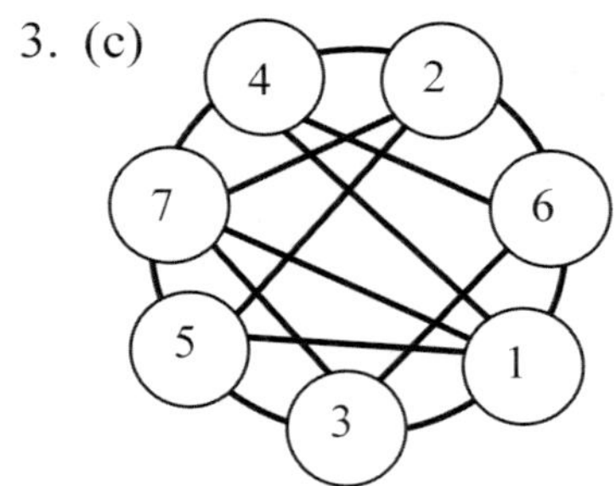

Number (6)

1. Boys: 4 Girls: 3
2. 28 seagulls
3. 309
4.
   2 x 5 cent, 20 x 2 cent, 15 x 10 cent
   4 x 5 cent, 40 x 2 cent, 10 x 10 cent
   6 x 5 cent, 60 x 2 cent, 5 x 10 cent
5. (a) (i) Original number: 12, number remaining: 3
   (ii) Original number: 8, number remaining: 2
   (iii) Original number: 6, number remaining: 2
   (b) Original number 108, remaining number is the cube of 3 (27).

Number (7)

1. (a) $a = 2$ (b) $a = 3$ (c) $a = 4$
2. (a)(i) $9 \times 0.4 = 3.6\text{m}$
   (ii) $11 \times 0.4 = 4.4\text{m}$
   (iii) $0.4(x + 1)$
   (b)(i) $9 \times 2 \times 0.4 + 8 \times 0.3 = 8.1\text{m}$
   (ii) $11 \times 2 \times 0.4 + 10 \times 0.3 = 11.8\text{m}$
   (iii) $2 \times 0.4(x + 1) + 0.3x = 1.1x + 0.8$
3. $2^{60} = 8^{20} = 32^{12}$, $4^{24} = 16^{12} = 64^{8}$
4. (a) 1000 metres (b) 3:2

Speed

1. $2\frac{1}{2}\text{km}$
2. Andrew walks at 3km/h and runs at 9km/h .
3. 30 minutes
4. The speeds of the cars are 59km/h and 85km/h .

Distance and Travel

1. (a) (i) 875km (ii) 735km
   (iii) 493km (iv) 417km
   (v) 118km (vi) 500km
   (vii) 502km (viii) 468km
   (b) 1 468km
   (c) 1 607km

1. (d)

	Adelaide	Coonalpyn	Bordertown	Nhill	Stawell	Ballarat	Melbourne
Melbourne	735	572	461	377	235	114	*
Ballarat	621	458	347	263	121	*	
Stawell	500	337	226	142	*		
Nhill	358	195	84	*			
Bordertown	274	111	*				
Coonalpyn	163	*					
Adelaide	*						

	Sydney	Mittagong	Goulburn	Gundagai	Wangaratta	Euroa	Melbourne
Melbourne	875	766	680	493	232	150	*
Euroa	725	616	530	343	82	*	
Wangaratta	643	534	448	261	*		
Gundagai	382	273	187	*			
Goulburn	195	86	*				
Mittagong	109	*					
Sydney	*						

2. One Solution: Jenny (54kg) and William (45kg) cross the river first (1st crossing) and Jenny paddles back leaving William on the other side (2nd crossing). Ian (85kg) crosses the river (85kg) (3rd crossing) and William paddles back (4th crossing). Jenny (54kg) and Lucinda (40 kg) paddle across (5th crossing) and Jenny returns to the other side again (6th crossing). Jenny picks up William and they join the others on the other side of the river (7th crossing).

   There are number of variations possible – Jenny, Lucinda and William are all interchangeable, only Ian's crossing is specific as his weight factor is important.

Number and Algebra (1)

1.

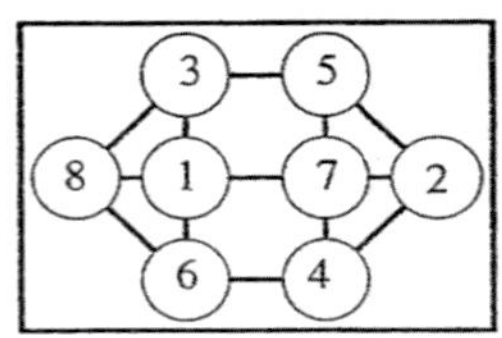

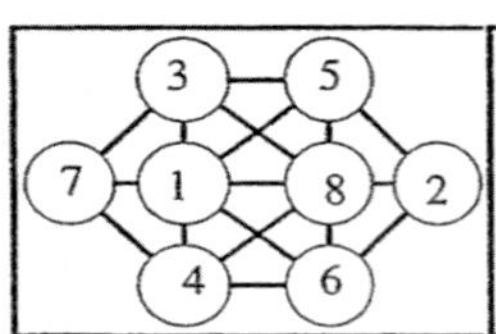

2. 10 bulls, 4 hens

3. Lucinda: 898 ± 25 → 873, 923
Jenny: 890 ± 33 → 857, 923
William: 950 ± 27 → 923, 977
Answer is 923 as this is the common value.

4 (a) 5 points 6 points

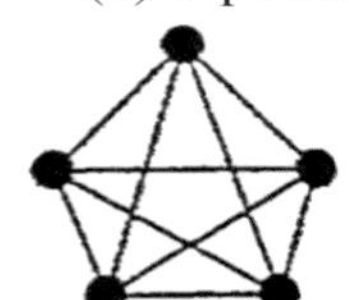

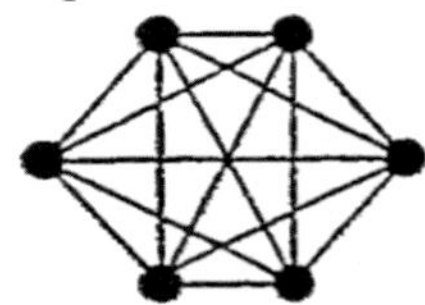

(b) This pattern operates just like a cumulative frequency distribution:
1 + 2 =3, 3 + 3 = 6, 6 + 4 = 10,
10 + 5 = 15, 15 + 6 = 21

(c) $L = \frac{P(P-1)}{2}$

(d) (i) 4 950 (ii) 499 500

No. of points	Lines
2	1
3	3
4	6
5	10
6	15
7	21
8	28
9	36
10	45

Number and Algebra (2)

1. {91, 93, 95, 97, 99}

2. 83

3. (a) $16.25
(b) (i) $69.50
(ii) $C = 1.3R + 1.45b$
(iii) 1 000 red, 500 blue, $2 025

4. (a) (i) 1 (ii) 2 (iii) 3
(b)

Row No. (n)	No. of dots (d)
1	1
2	2
3	3
4	4
5	5
6	6
7	7
8	8
9	9
10	10

(c) $n = d$
(d)

Pattern No. (P)	No. of rows	No. of dots (d)
1	1	1
2	2	3
3	3	6
4	4	10
5	5	15
6	6	21
7	7	27
8	8	36
9	9	45
10	10	55

(e) $d = \frac{P(P+1)}{2}$

(f) (i) $d = 5\ 050$ (ii) $d = 500\ 500$

Algebra

1. (a) 34, 36,38 (b) 43, 45
2. 3, 5, 7
3. 81kg
4.

Value of a	Value of b
7	7
4	8
3	9
2	12

5.

a	b	c
24	8	−12
13	16	−9
2	24	−6
−9	32	−3
−31	48	3

Shapes (1)

1. (a) (i) 1square (ii) 5 squares
   (iii) 14 squares
   (b) (i) 30 squares (ii) 55 squares
2. (a)

Size of board	Number of squares
1×1	1
2×2	5
3×3	14
4×4	30
5×5	55
6×6	91
7×7	140
8×8	204

(b)
$n^2 + (n-1)^2 + (n-2)^2 + ... + 3^2 + 2^2 + 1^2$

Shapes (2)

1. (a) (ii) 49 (iii) 36 (b) (i) 25 (ii) 16
2. (a) 24 (b) 25 (c) 18 (d) 18
3. (a) $N = n^2$, for $n \geq 1$
   (b) $N = (n-1)^2$, for $n \geq 2$
   (c) $N = (n-2)^2$, for $n \geq 3$

4. (a) (i)

Board size ($n \times n$)	Number of times shape fits on board N
2×2	0
3×3	2
4×4	4
5×5	8
6×6	12
7×7	18
8×8	24

(b) (i)

Board size ($n \times n$)	Number of times shape fits on board N
2×2	1
3×3	2
4×4	5
5×5	8
6×6	13
7×7	18
8×8	25

(c) (i)

Board size ($n \times n$)	Number of times shape fits on board N
3×3	1
4×4	2
5×5	5
6×6	8
7×7	13
8×8	18

(d) (i)

Board size ($n \times n$)	Number of times shape fits on board N
3×3	0
4×4	2
5×5	4
6×6	8
7×7	12
8×8	18

Space

1.

R	*B*
Y	*R*

R	*Y*
B	*R*

B	*R*
R	*Y*

Y	*R*
R	*B*

2.

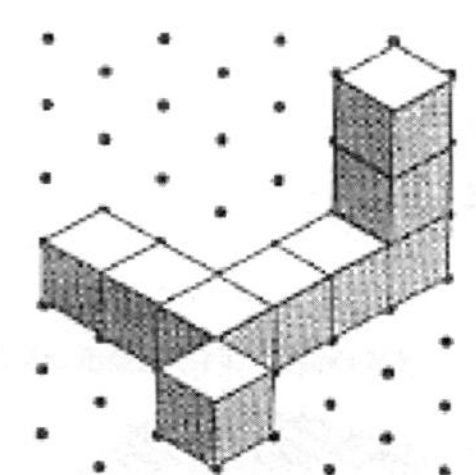

3.

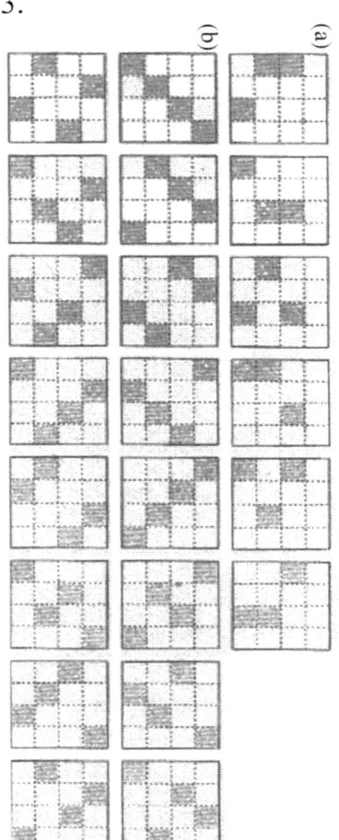

Measurement (1)

1. (a) $A: 4\text{cm}$, $B: 8\text{cm}$, $C: 12\text{cm}$

(b) $A: r = \dfrac{4}{2\pi} = \dfrac{2}{\pi}\text{cm}$

$B: r = \dfrac{8}{2\pi} = \dfrac{4}{\pi}\text{cm}$

$C: r = \dfrac{12}{2\pi} = \dfrac{6}{\pi}\text{cm}$

(c) (i) $r = \dfrac{4L}{2\pi} = \dfrac{2L}{\pi}$ (ii) $r \approx 6.37\text{cm}$

(d) (i) $r = \dfrac{L}{\sqrt{\pi}}$ (ii) $r \approx 1.13\text{cm}$

2. (a) $r = \dfrac{9}{2\pi}$

(b) $r = \dfrac{3L}{2\pi}$

(c) $r \approx 4.77\text{cm}$

Measurement (2)

1. 34

2. (a) πr (b) πr (c) πr

3. (a) 48 sqr. units (b) $25\pi - 48$

4. (a)396 sqr. units (b) 400 sqr. units

(c) The height of the smaller triangle on the top of the rectangle is not in the correct proportion to make a single large triangle.

(d) The height of the top triangle: $22\frac{1}{2}$ units : Area of large triangle:

$\frac{1}{2} \times 36 \times 22\frac{1}{2} = 405$

Area of parts:

$10 \times 20 + 2(\frac{1}{2} \times 8 \times 10) + 2(\frac{1}{2} \times 12.5 \times 10)$

$= 405$ sqr units